# 2005 EDITION

**ANSI/AF&PA SDPWS-2005**

**Approval Date: OCTOBER 26, 2005**

# ASD/LRFD

# WIND & SEISMIC

## SPECIAL DESIGN PROVISIONS FOR WIND AND SEISMIC

## WITH COMMENTARY

**Special Design Provisions for Wind and Seismic with Commentary 2005 Edition**

First Printing: August 2006
Second Printing: February 2007
Third Printing: February 2008

ISBN 0-9625985-3-4 (Volume 2)
ISBN 0-9625985-8-5 (4 Volume Set)



Copyright Permission
AF&PA American Wood Council
1111 Nineteenth St., NW, Suite 800
Washington, DC 20036
email: awcinfo@afandpa.org

Printed in the United States of America

# TABLE OF CONTENTS



# LIST OF TABLES

## LIST OF FIGURES

# DESIGNER FLOWCHART

AMERICAN FOREST & PAPER ASSOCIATION

# 1.1 Flowchart

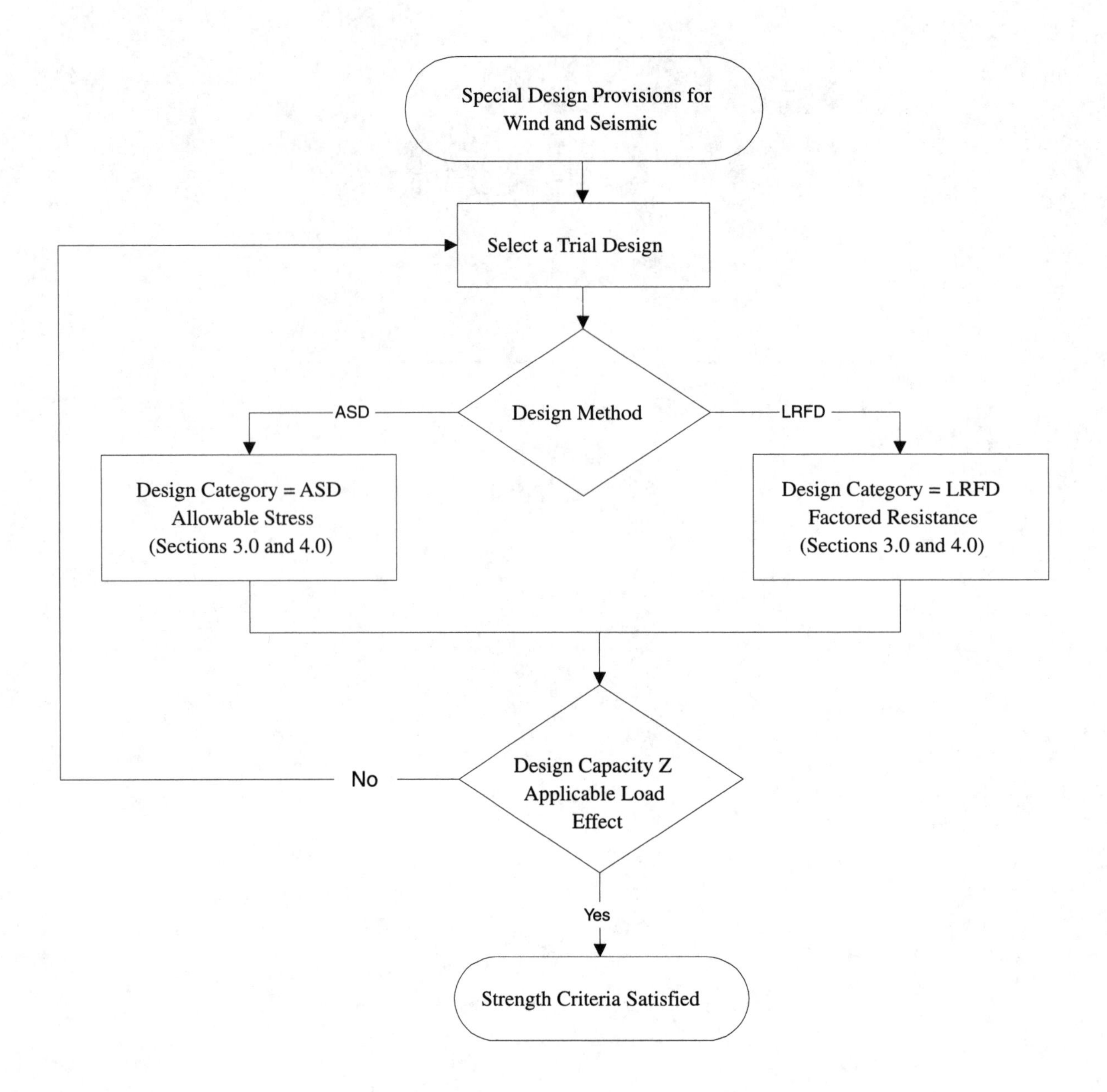

# GENERAL DESIGN REQUIREMENTS

# 2.1 General

## 2.1.1 Scope

The provisions of this document cover materials, design and construction of wood members, fasteners, and assemblies to resist wind and seismic forces.

## 2.1.2 Design Methods

Engineered design of wood structures to resist wind and seismic forces shall be by one of the methods described in 2.1.2.1 and 2.1.2.2.

> **Exception:** Wood structures shall be permitted to be constructed in accordance with prescriptive provisions permitted by the authority having jurisdiction.

2.1.2.1 Allowable Stress Design: Allowable stress design (ASD) shall be in accordance with the *National Design Specification® (NDS®) for Wood Construction* (ANSI/AF&PA NDS-05) and provisions of this document.

2.1.2.2 Strength Design: Load and resistance factor design (LRFD) of wood structures shall be in accordance with the *National Design Specification (NDS) for Wood Construction* (ANSI/AF&PA NDS-05) and provisions of this document.

# 2.2 Terminology

**ALLOWABLE STRESS DESIGN**. A method of proportioning structural members and their connections such that stresses do not exceed specified allowable stresses when the structure is subjected to appropriate load combinations (also called working stress design).

**ASD REDUCTION FACTOR**. A factor to reduce nominal strength to an allowable stress design level.

**BOUNDARY ELEMENT**. Diaphragm and shear wall boundary members to which sheathing transfers forces. Boundary elements include chords and collectors at diaphragm and shear wall perimeters, interior openings, discontinuities, and re-entrant corners.

**CHORD**. A boundary element perpendicular to the applied load that resists axial stresses due to the induced moment.

**COLLECTOR**. A diaphragm or shear wall element parallel and in line with the applied force that collects and transfers diaphragm shear forces to the vertical elements of the lateral-force-resisting system and/or distributes forces within the diaphragm.

**DIAPHRAGM**. A roof, floor, or other membrane bracing system acting to transmit lateral forces to the vertical resisting elements. When the term "diaphragm" is used, it includes horizontal bracing systems.

**DIAPHRAGM, BLOCKED**. A diaphragm in which all adjacent panel edges are fastened to either common framing or common blocking.

**DIAPHRAGM, FLEXIBLE**. A diaphragm is flexible for the purpose of distribution of story shear when the computed maximum in-plane deflection of the diaphragm itself under lateral load is greater than two times the average deflection of adjoining vertical elements of the lateral force resisting system of the associated story under equivalent tributary lateral load.

**DIAPHRAGM, RIGID**. A diaphragm is rigid for the purpose of distribution of story shear and torsional moment when the computed maximum in-plane deflection of the diaphragm itself under lateral load is less than or equal to two times the average deflection of adjoining vertical elements of the lateral force resisting system of the associated story under equivalent tributary lateral load. For analysis purposes, it can be assumed that a rigid diaphragm distributes story shear and torsional moment into lines of shear walls by the relative lateral stiffness of the shear walls.

**DIAPHRAGM BOUNDARY**. A location where shear is transferred into or out of the diaphragm sheathing. Transfer is either to a boundary element or to another force-resisting element.

**DIAPHRAGM, UNBLOCKED**. A diaphragm that has edge fasteners at supporting members only. Blocking between supporting structural members at panel edges is not included.

**FIBERBOARD**. A fibrous, homogeneous panel made from lignocellulosic fibers (usually wood or cane) and having a density of less than 31 pounds per cubic foot but more than 10 pounds per cubic foot.

**HARDBOARD**. A fibrous-felted, homogeneous panel made from lignocellulosic fibers consolidated under heat and pressure in a hot press to a density not less than 31 pounds per cubic foot.

**LATERAL STIFFNESS**. The inverse of the deformation of shear walls under an applied unit load, or the force required to deform a shear wall a unit distance.

**LOAD AND RESISTANCE FACTOR DESIGN (LRFD)**. A method of proportioning structural members and their connections using load and resistance factors such that no applicable limit state is reached when the structure is subjected to appropriate load combinations.

**NOMINAL STRENGTH**. Strength of a member, cross section, or connection before application of any strength reduction factors.

**ORIENTED STRAND BOARD (OSB)**. A mat-formed wood structural panel product composed of thin rectangular wood strands or wafers arranged in oriented layers and bonded with waterproof adhesive.

**PARTICLEBOARD**. A generic term for a panel primarily composed of cellulosic materials (usually wood), generally in the form of discrete pieces or particles, as distinguished from fibers. The cellulosic material is combined with synthetic resin or other suitable bonding system by a process in which the interparticle bond is created by the bonding system under heat and pressure.

**PERFORATED SHEAR WALL**. A sheathed wall with openings, but which has not been specifically designed and detailed for force transfer around wall openings.

**PERFORATED SHEAR WALL SEGMENT**. A section of a perforated shear wall with full height sheathing that meets the requirements for maximum aspect ratio limits in 4.3.4.

**PLYWOOD.** A wood structural panel comprised of plies of wood veneer arranged in cross-aligned layers. The plies are bonded with an adhesive that cures on application of heat and pressure.

**REQUIRED STRENGTH**. Strength of a member, cross section, or connection required to resist factored loads or related internal moments and forces.

**RESISTANCE FACTOR**. A factor that accounts for deviations of the actual strength from the nominal strength and the manner and consequences of failure.

**SEISMIC DESIGN CATEGORY**. A classification assigned to a structure based on its Seismic Use Group (see building code) and the severity of the design earthquake ground motion at the site.

**SHEAR WALL**. A wall designed to resist lateral forces parallel to the plane of a wall.

**SHEAR WALL LINE**. A series of shear walls in a line at a given story level.

**SUBDIAPHRAGM**. A portion of a larger wood diaphragm designed to anchor and transfer local forces to primary diaphragm struts and the main diaphragm.

**TIE-DOWN (HOLD DOWN)**. A device used to resist uplift of the chords of shear walls.

**WOOD STRUCTURAL PANEL**. A panel manufactured from veneers; or wood strands or wafers; or a combination of veneer and wood strands or wafers; bonded together with waterproof synthetic resins or other suitable bonding systems. Examples of wood structural panels are plywood, oriented strand board (OSB), or composite panels.



## 2.3 Notation

A = area, in.$^2$

C = compression chord force, lbs

$C_o$ = shear capacity adjustment factor

E = modulus of elasticity, psi

G = specific gravity

$G_a$ = apparent shear stiffness from nail slip and panel shear deformation, kips/in.

$G_{ac}$ = combined apparent shear wall shear stiffness of two-sided shear wall, kips/in.

$G_{a1}$ = apparent shear wall shear stiffness for side 1, kips/in.

$G_{a2}$ = apparent shear wall shear stiffness for side 2, kips/in.

$K_{min}$ = minimum ratio of $\nu_1/G_{a1}$ or $\nu_2/G_{a2}$

L = dimension of a diaphragm in the direction perpendicular to the application of force and is measured as the distance between vertical elements of the lateral force-resisting system (in many cases, this will match the sheathed dimensions), ft. For open front structures, L is the length from the edge of the diaphragm at the open front to the vertical-resisting elements parallel to the direction of the applied force, ft

$L_c$ = length of the cantilever for a cantilever diaphragm, ft

$\Sigma L_i$ = sum of perforated shear wall segment lengths, ft

R = response modification coefficient

T = tension chord force, lbs

V = induced shear force in perforated shear wall, lbs

W = dimension of a diaphragm in the direction of application of force and is measured as the distance between diaphragm chords, ft (in many cases, this will match the sheathed dimension)

b = length of a shear wall or shear wall segment measured as the sheathed dimension of the shear wall or segment, ft

$b_s$ = shear wall length for determining aspect ratio. For perforated shear walls, use the minimum shear wall segment length included in the $\Sigma L_i$, ft

h = height of a shear wall or shear wall segment, ft, measured as:

1. maximum clear height from top of foundation to bottom of diaphragm framing above, ft, or
2. maximum clear height from top of diaphragm below to bottom of diaphragm framing above, ft

t = uniform uplift force, lbs/ft

$\nu$ = induced unit shear, lbs/ft

$\nu_{max}$ = maximum induced unit shear force, lbs/ft

$\nu_s$ = nominal unit shear capacity for seismic design, lbs/ft

$\nu_{sc}$ = combined nominal unit shear capacity of two-sided shear wall for seismic design, lbs/ft

$\nu_{s1}$ = nominal unit shear capacity for designated side 1, lbs/ft

$\nu_{s2}$ = nominal unit shear capacity for designated side 2, lbs/ft

$\nu_w$ = nominal unit shear capacity for wind design, lbs/ft

$\nu_{wc}$ = combined nominal unit shear capacity of two-sided shear wall for wind design, lbs/ft

x = distance from chord splice to nearest support, ft

$\Delta_a$ = total vertical elongation of wall anchorage system (including fastener slip, device elongation, rod elongation, etc.), in., at the induced unit shear in the shear wall

$\Delta_c$ = diaphragm chord splice slip at the induced unit shear in diaphragm, in.

$\delta_{dia}$ = maximum diaphragm deflection determined by elastic analysis, in.

$\delta_{sw}$ = maximum shear wall deflection determined by elastic analysis, in.

$\phi_b$ = sheathing resistance factor for out-of-plane bending

$\phi_D$ = sheathing resistance factor for in-plane shear of shear walls and diaphragms

$\Omega_0$ = system overstrength factor

# MEMBERS AND CONNECTIONS

# 3.1 Framing

## 3.1.1 Wall Framing

In addition to gravity loads, wall framing shall be designed to resist induced wind and seismic forces. The framing shall be designed using the methods referenced in 2.1.2.1 for allowable stress design (ASD) and 2.1.2.2 for strength design (LRFD).

3.1.1.1 Wall Stud Bending Stress Increase: The bending stress, $F_b$, for sawn lumber wood studs resisting out-of-plane wind loads shall be permitted to be increased by the factors in Table 3.1.1.1, in lieu of the 1.15 repetitive member factor, to take into consideration the load sharing and composite action provided by wood structural panel sheathing. The factor applies when studs are designed for bending, spaced no more than 16" on center, covered on the inside with a minimum of ½" gypsum wallboard, attached in accordance with minimum building code requirements and sheathed on the exterior with a minimum of 3/8" wood structural panel sheathing with all panel joints occurring over studs or blocking and attached using a minimum of 8d common nails spaced a maximum of 6" on center at panel edges and 12" on center at intermediate framing members.

**Table 3.1.1.1 Wall Stud Bending Stress Increase Factors**

| Stud Size | System Factor |
|---|---|
| 2x4 | 1.50 |
| 2x6 | 1.35 |
| 2x8 | 1.25 |
| 2x10 | 1.20 |
| 2x12 | 1.15 |

## 3.1.2 Floor Framing

In addition to gravity loads, floor framing shall be designed to resist induced wind and seismic forces. The framing shall be designed using the methods referenced in 2.1.2.1 for allowable stress design (ASD) and 2.1.2.2 for strength design (LRFD).

## 3.1.3 Roof Framing

In addition to gravity loads, roof framing shall be designed to resist induced wind and seismic forces. The framing shall be designed using the methods referenced in 2.1.2.1 for allowable stress design (ASD) and 2.1.2.2 for strength design (LRFD).

# 3.2 Sheathing

## 3.2.1 Wall Sheathing

Exterior wall sheathing and its fasteners shall be capable of resisting and transferring wind loads to the wall framing. Maximum spans and nominal uniform load capacities for wall sheathing materials are given in Table 3.2.1. The ASD allowable uniform load capacities to be used for wind design shall be determined by dividing the nominal uniform load capacities in Table 3.2.1 by an ASD reduction factor of 1.6. The LRFD factored uniform load capacities to be used for wind design shall be determined by multiplying the nominal uniform load capacities in Table 3.2.1 by a resistance factor, $\phi_b$, of 0.85. Sheathing used in shear wall assemblies to resist lateral forces shall be designed in accordance with 4.3.

## Table 3.2.1 Nominal Uniform Load Capacities (psf) for Wall Sheathing Resisting Out-of-Plane Wind Loads[1]

| Sheathing Type[3] | Span Rating or Grade | Minimum Thickness (in.) | Strength Axis[5]: Perpendicular to Supports: Maximum Stud Spacing (in.) | Perpendicular to Supports: Actual Stud Spacing (in.) 12 — Nominal Uniform Loads (psf) | Perpendicular: 16 | Perpendicular: 24 | Parallel to Supports: Maximum Stud Spacing (in.) | Parallel to Supports: Actual Stud Spacing (in.) 12 — Nominal Uniform Loads (psf) | Parallel: 16 | Parallel: 24 |
|---|---|---|---|---|---|---|---|---|---|---|
| Wood Structural Panels (Sheathing Grades, C-C, C-D, C-C Plugged, OSB)[4] | 24/0 | 3/8 | 24 | 425 | 240 | 105 | 24 | 90 | 50 | 25[2] |
| | 24/16 | 7/16 | 24 | 540 | 305 | 135 | 24 | 110 | 60 | 25[2] |
| | 32/16 | 15/32 | 24 | 625 | 355 | 155 | 24 | 155 | 90 | 40[2] |
| | 40/20 | 19/32 | 24 | 955 | 595 | 265 | 24 | 255 | 145 | 65[2] |
| | 48/24 | 23/32 | 24 | 1160 | 805 | 360 | 24 | 380 | 215 | 95[2] |
| Particleboard Sheathing (M-S Exterior Glue) | | 3/8 | 16 | (contact manufacturer) | | | 16 | (contact manufacturer) | | |
| | | 1/2 | 16 | | | | 16 | | | |
| Particleboard Panel Siding (M-S Exterior Glue) | | 5/8 | 16 | (contact manufacturer) | | | 16 | (contact manufacturer) | | |
| | | 3/4 | 24 | | | | 24 | | | |
| Hardboard Siding (Direct to Studs) | Lap Siding | 7/16 | 16 | 460 | 260 | - | - | - | - | - |
| | Shiplap Edge Panel Siding | 7/16 | 24 | 460 | 260 | 115 | 24 | 460 | 260 | 115 |
| | Square Edge Panel Siding | 7/16 | 24 | 460 | 260 | 115 | 24 | 460 | 260 | 115 |
| Cellulosic Fiberboard Sheathing | Regular | 1/2 | 16 | 90 | 50 | - | 16 | 90 | 50 | - |
| | Structural | 1/2 | 16 | 135 | 75 | - | 16 | 135 | 75 | - |
| | Structural | 25/32 | 16 | 165 | 90 | - | 16 | 165 | 90 | - |

1. Nominal capacities shall be adjusted in accordance with Section 3.2.1 to determine ASD uniform load capacity and LRFD uniform resistances.
2. Sheathing shall be plywood with 4 or more plies or OSB.
3. Wood structural panels shall conform to the requirements for its type in DOC PS 1 or PS 2. Particleboard sheathing shall conform to ANSI A208.1. Hardboard panel and siding shall conform to the requirements of AHA A135.5 or AHA A135.4 as applicable. Cellulosic fiberboard sheathing shall conform to AHA A194.1 or ASTM C 208.
4. Tabulated values are for maximum bending loads from wind. Loads are limited by bending or shear stress assuming a two-span continuous condition. Where panels are continuous over 3 or more spans, the tabulated values shall be permitted to be increased in accordance with the *ASD/LRFD Manual for Engineered Wood Construction*.
5. Strength axis is defined as the axis parallel to the face and back orientation of the flakes or the grain (veneer), which is generally the long panel direction, unless otherwise marked.

## 3.2.2 Floor Sheathing

Floor sheathing shall be capable of resisting and transferring gravity loads to the floor framing. Sheathing used in diaphragm assemblies to resist lateral forces shall be designed in accordance with 4.2.

## 3.2.3 Roof Sheathing

Roof sheathing and its fasteners shall be capable of resisting and transferring wind and gravity loads to the roof framing. Maximum spans and nominal uniform load capacities for roof sheathing materials are given in Table 3.2.2. The ASD allowable uniform load capacities to be used for wind design shall be determined by dividing the nominal uniform load capacities in Table 3.2.2 by an ASD reduction factor of 1.6. The LRFD factored uniform load capacities to be used for wind design shall be determined by multiplying the nominal uniform load capacities in Table 3.2.2 by a resistance factor, $\phi_b$, of 0.85. Sheathing used in diaphragm assemblies to resist lateral forces shall be designed in accordance with 4.2.

**Table 3.2.2 Nominal Uniform Load Capacities (psf) for Roof Sheathing Resisting Out-of-Plane Wind Loads[1,3]**

| Sheathing Type[2] | Span Rating or Grade | Minimum Thickness (in.) | Strength Axis[4] Applied Perpendicular to Supports | | | | | |
|---|---|---|---|---|---|---|---|---|
| | | | Rafter/Truss Spacing (in.) | | | | | |
| | | | 12 | 16 | 19.2 | 24 | 32 | 48 |
| | | | Nominal Uniform Loads (psf) | | | | | |
| Wood Structural Panels (Sheathing Grades, C-C, C-D, C-C Plugged, OSB) | 24/0 | 3/8 | 425 | 240 | 165 | 105 | - | - |
| | 24/16 | 7/16 | 540 | 305 | 210 | 135 | - | - |
| | 32/16 | 15/32 | 625 | 355 | 245 | 155 | 90 | - |
| | 40/20 | 19/32 | 955 | 595 | 415 | 265 | 150 | - |
| | 48/24 | 23/32 | 1160 | 805 | 560 | 360 | 200 | 90 |
| Wood Structural Panels (Single Floor Grades, Underlayment, C-C Plugged) | 16 o.c. | 19/32 | 705 | 395 | 275 | 175 | 100 | - |
| | 20 o.c. | 19/32 | 815 | 455 | 320 | 205 | 115 | - |
| | 24 o.c. | 23/32 | 1085 | 610 | 425 | 270 | 150 | - |
| | 32 o.c. | 7/8 | 1395 | 830 | 575 | 370 | 205 | 90 |
| | 48 o.c. | 1-1/8 | 1790 | 1295 | 1060 | 680 | 380 | 170 |

1. Nominal capacities shall be adjusted in accordance with Section 3.2.3 to determine ASD uniform load capacity and LRFD uniform resistances.
2. Wood structural panels shall conform to the requirements for its type in DOC PS 1 or PS 2.
3. Tabulated values are for maximum bending loads from wind. Loads are limited by bending or shear stress assuming a two-span continuous condition. Where panels are continuous over 3 or more spans, the tabulated values shall be permitted to be increased in accordance with the *ASD/LRFD Manual for Engineered Wood Construction.*
4. Strength axis is defined as the axis parallel to the face and back orientation of the flakes or the grain (veneer), which is generally the long panel direction, unless otherwise marked.

# 3.3 Connections

Connections resisting induced wind and seismic forces shall be designed in accordance with the methods referenced in 2.1.2.1 for allowable stress design (ASD) and 2.1.2.2 for strength design (LRFD).

# LATERAL FORCE-RESISTING SYSTEMS



4

# 4.1 General

## 4.1.1 Design Requirements

The proportioning, design, and detailing of engineered wood systems, members, and connections in lateral force-resisting systems shall be in accordance with the reference documents in 2.1.2 and provisions in this chapter. A continuous load path, or paths, with adequate strength and stiffness shall be provided to transfer all forces from the point of application to the final point of resistance.

## 4.1.2 Shear Capacity

Nominal shear capacities of diaphragms and shear walls are provided for reference assemblies in Tables 4.2A, 4.2B, and 4.2C and Tables 4.3A, 4.3B, and 4.3C, respectively. Alternatively, shear capacity of diaphragms and shear walls shall be permitted to be calculated by principles of mechanics using values of fastener strength and sheathing shear capacity.

## 4.1.3 Deformation Requirements

Deformation of connections within and between structural elements shall be considered in design such that the deformation of each element and connection comprising the lateral force-resisting system is compatible with the deformations of the other lateral force-resisting elements and connections and with the overall system.

## 4.1.4 Boundary Elements

Shear wall and diaphragm boundary elements shall be provided to transfer the design tension and compression forces. Diaphragm and shear wall sheathing shall not be used to splice boundary elements. Diaphragm chords and collectors shall be placed in, or in contact with, the plane of the diaphragm framing unless it can be demonstrated that the moments, shears, and deflections, considering eccentricities resulting from other configurations, can be tolerated without exceeding the framing capacity and drift limits.

## 4.1.5 Wood Members and Systems Resisting Seismic Forces Contributed by Masonry and Concrete Walls

Wood shear walls, diaphragms, trusses, and other wood members and systems shall not be used to resist seismic forces contributed by masonry or concrete walls in structures over one story in height.

**Exceptions:**

1. Wood floor and roof members shall be permitted to be used in diaphragms and horizontal trusses to resist horizontal seismic forces contributed by masonry or concrete walls provided such forces do not result in torsional force distribution through the diaphragm or truss.
2. Vertical wood structural panel sheathed shear walls shall be permitted to be used to provide resistance to seismic forces in two-story structures of masonry or concrete walls, provided the following requirements are met:
   a. Story-to-story wall heights shall not exceed 12'.
   b. Diaphragms shall not be considered to transmit lateral forces by torsional force distribution or cantilever past the outermost supporting shear wall.
   c. Combined deflections of diaphragms and shear walls shall not permit per story drift of supported masonry or concrete walls to exceed 0.7% of the story height.
   d. Wood structural panel sheathing in diaphragms shall have all unsupported edges blocked. Wood structural panel sheathing for both stories of shear walls shall have all unsupported edges blocked and, for the lower story, shall have a minimum thickness of 15/32".
   e. There shall be no out-of-plane horizontal offsets between the first and second stories of wood structural panel shear walls.

### 4.1.6 Wood Members and Systems Resisting Seismic Forces from Other Concrete or Masonry Construction

Wood members and systems shall be designed to resist seismic forces from other concrete or masonry components, including but not limited to: chimneys, fireplaces, concrete or masonry veneers, and concrete floors.

### 4.1.7 Toe-Nailed Connections

In seismic design categories D, E, and F, toe-nailed connections shall not be used to transfer seismic lateral forces greater than 150 pounds per lineal foot for ASD and 205 pounds per lineal foot for LRFD from diaphragms to shear walls, collectors, or other elements, or from shear walls to other elements.





## 4.2 Wood Diaphragms

### 4.2.1 Application Requirements

Wood diaphragms are permitted to be used to resist lateral forces provided the deflection in the plane of the diaphragm, as determined by calculations, tests, or analogies drawn therefrom, does not exceed the maximum permissible deflection limit of attached load distributing or resisting elements. Permissible deflection shall be that deflection that will permit the diaphragm and any attached elements to maintain their structural integrity and continue to support their prescribed loads as determined by the applicable building code or standard. Connections and blocking shall extend into the diaphragm a sufficient distance to develop the force transferred into the diaphragm.

### 4.2.2 Deflection

Calculations of diaphragm deflection shall account for bending and shear deflections, fastener deformation, chord splice slip, and other contributing sources of deflection.

The diaphragm deflection, $\delta_{dia}$, is permitted to be calculated by use of the following equation:

$$\delta_{dia} = \frac{5\nu L^3}{8EAW} + \frac{0.25\nu L}{1000G_a} + \frac{\sum(x\Delta_c)}{2W} \qquad (4.2\text{-}1)$$

**where:**

$E$ = modulus of elasticity of diaphragm chords, psi

$A$ = area of chord cross-section, in.$^2$

$G_a$ = apparent diaphragm shear stiffness from nail slip and panel shear deformation, kips/in. (from Column A, Tables 4.2A, 4.2B, or 4.2C)

$L$ = diaphragm length, ft

$\nu$ = induced unit shear in diaphragm, lbs/ft

$W$ = diaphragm width, ft

$x$ = distance from chord splice to nearest support, ft

$\Delta_c$ = diaphragm chord splice slip, in., at the induced unit shear in diaphragm

$\delta_{dia}$ = maximum mid-span diaphragm deflection determined by elastic analysis, in.

Alternatively, for wood structural panel diaphragms, deflection is permitted to be calculated using a rational analysis where apparent shear stiffness accounts for panel shear deformation and non-linear nail slip in the sheathing-to-framing connection.

### 4.2.3 Unit Shear Capacities

The nominal unit shear capacities for seismic design are provided in Column A of Tables 4.2A, 4.2B, and 4.2C; and for wind design in Column B of Tables 4.2A, 4.2B, and 4.2C. The ASD allowable unit shear capacity shall be determined by dividing the nominal unit shear capacity by the ASD reduction factor of 2.0. No further increases shall be permitted. The LRFD factored unit resistance shall be determined by multiplying the nominal unit shear capacity by a resistance factor, $\phi_D$, of 0.80.

## 4.2.4 Diaphragm Aspect Ratios

Size and shape of diaphragms shall be limited to the aspect ratios in Table 4.2.4.

**Table 4.2.4 Maximum Diaphragm Aspect Ratios**
(Horizontal or Sloped Diaphragms)

| Diaphragm Sheathing Type | Maximum L/W Ratio |
|---|---|
| Wood structural panel, unblocked | 3:1 |
| Wood structural panel, blocked | 4:1 |
| Single-layer straight lumber sheathing | 2:1 |
| Single-layer diagonal lumber sheathing | 3:1 |
| Double-layer diagonal lumber sheathing | 4:1 |

## 4.2.5 Horizontal Distribution of Shear

Diaphragms shall be defined as rigid or flexible for the purposes of distributing shear loads and designing for torsional moments. When a diaphragm is defined as flexible, the diaphragm shear forces shall be distributed to the vertical-resisting elements based on tributary area. When a diaphragm is defined as rigid, the diaphragm shear forces shall be distributed based on the relative lateral stiffnesses of the vertical-resisting elements of the story below.

4.2.5.1 Torsional Irregularity: Structures with rigid wood diaphragms shall be considered as torsionally irregular when the maximum story drift, computed including accidental torsion, at one end of the structure is more than 1.2 times the average of the story drifts at the two ends of the structure. Where torsional irregularity exists, diaphragms shall meet the following requirements:

1. The diaphragm conforms to 4.2.7.1 through 4.2.7.3.
2. The L/W ratio of the diaphragm is less than 1:1 for one-story structures or 1:1-½ for structures over one story in height.

**Exception:** Where calculations show that diaphragm deflections can be tolerated, the length, L, shall be permitted to be increased to an L/W ratio not greater than 1-½:1 when sheathed in conformance with 4.2.7.1, or to 1:1 when sheathed in conformance with 4.2.7.2 or 4.2.7.3.

4.2.5.1.1 Open Front Structures: Open front structures utilizing rigid wood diaphragms to distribute shear forces through torsion shall be permitted provided:

1. The diaphragm length, L, (normal to the open side) does not exceed 25'.
2. The L/W ratio (as shown in Figure 4A) of the diaphragm is less than 1:1 for one-story structures or 1:1-½ for structures over one story in height.

**Exception:** Where calculations show that diaphragm deflections can be tolerated, the length, L, (normal to the open side) shall be permitted to be increased to an L/W ratio not greater than 1-½:1 when sheathed in conformance with 4.2.7.1 or 4.2.7.3, or to 1:1 when sheathed in conformance with 4.2.7.2.

**Figure 4A Open Front Structure**

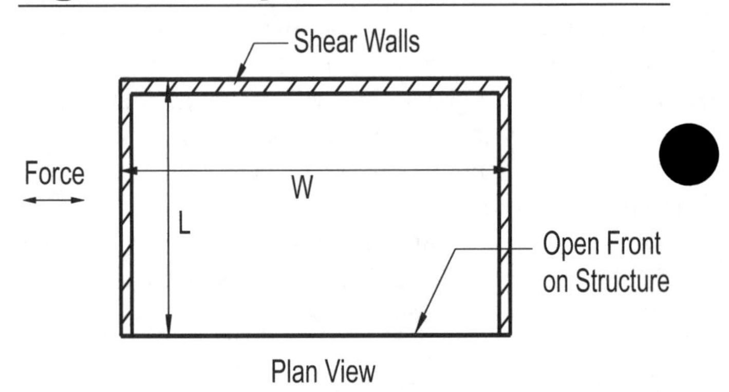


4.2.5.2 Cantilevered Diaphragms: Rigid wood diaphragms shall be permitted to cantilever past the outermost supporting shear wall (or other vertical resisting element) a distance, $L_c$, of not more than 25' or 2/3 of the diaphragm width, W, whichever is smaller. Figure 4B illustrates the dimensions of $L_c$ and W for a cantilevered diaphragm.

**Figure 4B Cantilevered Building**

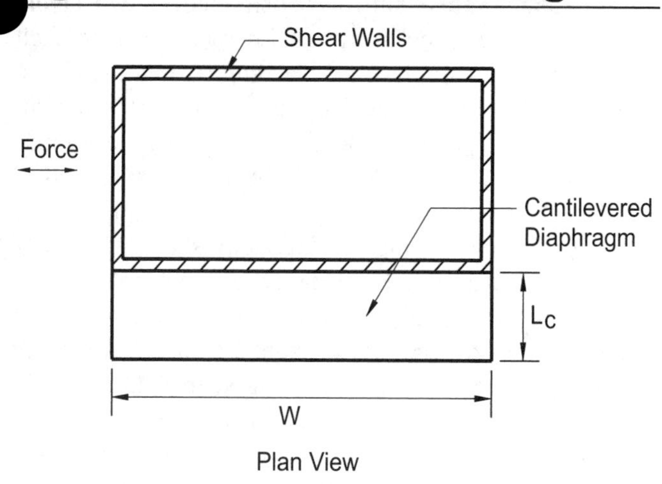


## 4.2.6 Construction Requirements

4.2.6.1 Framing Requirements: Diaphragm boundary elements shall be provided to transmit the design tension, compression, and shear forces. Diaphragm sheathing shall not be used to splice boundary elements. Diaphragm chords and collectors shall be placed in, or in contact with, the plane of the diaphragm framing unless it can be demonstrated that the moments, shears, and deflections, considering eccentricities resulting from other configurations, can be tolerated without exceeding the framing capacity and drift limits.

4.2.6.2 Sheathing: Diaphragms shall be sheathed with approved materials. Details on sheathing types and thicknesses for commonly used floor, roof, and ceiling diaphragm assemblies are provided in 4.2.7 and Tables 4.2A, 4.2B, and 4.2C.

4.2.6.3 Fasteners: Sheathing shall be attached to framing using fasteners alone, or in combination with adhesives. Nails or other approved sheathing fasteners shall be driven with the head of the fastener flush with the surface of the sheathing. Details on type, size, and spacing of mechanical fasteners for typical floor, roof, and ceiling diaphragm assemblies are provided in 4.2.7 and Tables 4.2A, 4.2B, and 4.2C.

## 4.2.7 Diaphragm Assemblies

4.2.7.1 Wood Structural Panel Diaphragms: Diaphragms sheathed with wood structural panel sheathing shall be permitted to be used to resist seismic and wind forces. Wood structural panel sheathing used for diaphragms that are part of the lateral force-resisting system shall be applied directly to the framing members.

**Exception:** Wood structural panel sheathing in a diaphragm is permitted to be fastened over solid lumber planking or laminated decking provided the following requirements are met:

1. Panel edges do not coincide with joints in the lumber planking or laminated decking.
2. Adjacent panel edges parallel to the planks or decking are fastened to a common member.
3. The planking or decking shall be of sufficient thickness to satisfy minimum fastener penetration in framing requirements as given in Table 4.2A.
4. Diaphragm aspect ratio (L/W) does not exceed that for a blocked wood structural panel diaphragm (4:1).
5. Diaphragm forces are transferred from wood structural panel sheathing to diaphragm boundary elements through planking or decking or by other methods.

4.2.7.1.1 Blocked and Unblocked Diaphragms: Where diaphragms are designated as blocked, all joints in sheathing shall occur over and be fastened to common framing members. The size and spacing of fasteners at wood diaphragm boundaries, panel edges, and intermediate supports shall be as prescribed in Tables 4.2A and 4.2B. The diaphragm shall be constructed as follows:

1. Panels not less than 4' x 8' except at boundaries and changes in framing where minimum panel dimension shall be 24" unless all edges of the undersized panels are supported by framing members or blocking.
2. Nails located at least 3/8" from edges and ends of panels. Maximum nail spacing of 6 inches on center at panel edges. Maximum nail spacing of 6" on center along intermediate framing members when supports are spaced 48" on center. Maximum nail spacing along intermediate framing of 12" on center for closer support spacings.
3. 2" nominal or wider framing thickness at adjoining panel edges except that 3" nominal or wider framing thickness and staggered nailing are required where:
   a. nails are spaced 2-½" on center or less at adjoining panel edges, or

b. 10d nails having penetration into framing of more than 1-5/8" are spaced 3" on center or less at adjoining panel edges.

4. Wood structural panels shall conform to the requirements for its type in DOC PS 1 or PS 2.

4.2.7.2 Diaphragms Diagonally Sheathed with Single-Layer of Lumber: Single diagonally sheathed lumber diaphragms are permitted to be used to resist seismic and wind forces. Single diagonally sheathed lumber diaphragms shall be constructed of minimum 1" thick nominal sheathing boards or 2" thick nominal lumber laid at an angle of approximately 45° to the supports. End joints in adjacent boards shall be separated by at least one joist space and there shall be at least two boards between joints on the same support. Nailing of diagonally sheathed lumber diaphragms shall be in accordance with Table 4.2C. Single diagonally sheathed lumber diaphragms shall be permitted to consist of 2" nominal lumber (1-½" thick) where the supports are not less than 3" nominal (2-½" thick) in width or 4" nominal (3-½" deep) in depth.

4.2.7.3 Diaphragms Diagonally Sheathed with Double-Layer of Lumber: Double diagonally sheathed lumber diaphragms are permitted to be used to resist seismic and wind forces. Double diagonally sheathed lumber diaphragms shall be constructed of two layers of diagonal sheathing boards laid perpendicular to each other on the same face of the supporting members. Each chord shall be considered as a beam with uniform load per foot equal to 50% of the unit shear due to diaphragm action. The load shall be assumed as acting normal to the chord in the plane of the diaphragm in either direction. Nailing of diagonally sheathed lumber diaphragms shall be in accordance with Table 4.2C.

4.2.7.4 Diaphragms Horizontally Sheathed with Single-Layer of Lumber: Horizontally sheathed lumber diaphragms are permitted to be used to resist seismic and wind forces. Horizontally sheathed lumber diaphragms shall be constructed of minimum 1" thick nominal sheathing boards or minimum 2" thick nominal lumber laid perpendicular to the supports. End joints in adjacent boards shall be separated by at least one joist space and there shall be at least two boards between joints on the same support. Nailing of horizontally sheathed lumber diaphragms shall be in accordance with Table 4.2C.

# Table 4.2A Nominal Unit Shear Capacities for Wood-Frame Diaphragms

## Blocked Wood Structural Panel Diaphragms (Excluding Plywood for $G_a$)[1,2,3,4]

| Sheathing Grade | Common Nail Size | Minimum Fastener Penetration in Framing (in.) | Minimum Nominal Panel Thickness (in.) | Minimum Nominal Framing Width (in.) | A SEISMIC | | | | | | | | B WIND | | | |
|---|---|---|---|---|---|---|---|---|---|---|---|---|---|---|---|---|
| | | | | | Nail Spacing (in.) at diaphragm boundaries (all cases), at continuous panel edges parallel to load (Cases 3 & 4), and at all panel edges (Cases 5 & 6) | | | | | | | | Nail Spacing (in.) at diaphragm boundaries (all cases), at continuous panel edges parallel to load (Cases 3 & 4), and at all panel edges (Cases 5 & 6) | | | |
| | | | | | 6 | | 4 | | 2-1/2 | | 2 | | 6 | 4 | 2-1/2 | 2 |
| | | | | | Nail Spacing (in.) at other panel edges (Cases 1, 2, 3, & 4) | | | | | | | | Nail Spacing (in.) at other panel edges (Cases 1, 2, 3, & 4) | | | |
| | | | | | 6 | | 6 | | 4 | | 3 | | 6 | 6 | 4 | 3 |
| | | | | | $v_s$ (plf) | $G_a$ (kips/in.) | $v_s$ (plf) | $G_a$ (kips/in.) | $v_s$ (plf) | $G_a$ (kips/in.) | $v_s$ (plf) | $G_a$ (kips/in.) | $v_w$ (plf) | $v_w$ (plf) | $v_w$ (plf) | $v_w$ (plf) |
| Structural I | 6d | 1-1/4 | 5/16 | 2 | 370 | 15.0 | 500 | 8.5 | 750 | 12.0 | 840 | 20.0 | 520 | 700 | 1050 | 1175 |
| | | | | 3 | 420 | 12.0 | 560 | 7.0 | 840 | 9.5 | 950 | 17.0 | 590 | 785 | 1175 | 1330 |
| | 8d | 1-3/8 | 3/8 | 2 | 540 | 14.0 | 720 | 9.0 | 1060 | 13.0 | 1200 | 21.0 | 755 | 1010 | 1485 | 1680 |
| | | | | 3 | 600 | 12.0 | 800 | 7.5 | 1200 | 10.0 | 1350 | 18.0 | 840 | 1120 | 1680 | 1890 |
| | 10d | 1-1/2 | 15/32 | 2 | 640 | 24.0 | 850 | 15.0 | 1280 | 20.0 | 1460 | 31.0 | 895 | 1190 | 1790 | 2045 |
| | | | | 3 | 720 | 20.0 | 960 | 12.0 | 1440 | 16.0 | 1640 | 26.0 | 1010 | 1345 | 2015 | 2295 |
| Sheathing and Single-Floor | 6d | 1-1/4 | 5/16 | 2 | 340 | 15.0 | 450 | 9.0 | 670 | 13.0 | 760 | 21.0 | 475 | 630 | 940 | 1065 |
| | | | | 3 | 380 | 12.0 | 500 | 7.0 | 760 | 10.0 | 860 | 17.0 | 530 | 700 | 1065 | 1205 |
| | | | 3/8 | 2 | 370 | 13.0 | 500 | 7.0 | 750 | 10.0 | 840 | 18.0 | 520 | 700 | 1050 | 1175 |
| | | | | 3 | 420 | 10.0 | 560 | 5.5 | 840 | 8.5 | 950 | 14.0 | 590 | 785 | 1175 | 1330 |
| | 8d | 1-3/8 | 3/8 | 2 | 480 | 15.0 | 640 | 9.5 | 960 | 13.0 | 1090 | 21.0 | 670 | 895 | 1345 | 1525 |
| | | | | 3 | 540 | 12.0 | 720 | 7.5 | 1080 | 11.0 | 1220 | 18.0 | 755 | 1010 | 1510 | 1710 |
| | | | 7/16 | 2 | 510 | 14.0 | 680 | 8.5 | 1010 | 12.0 | 1150 | 20.0 | 715 | 950 | 1415 | 1610 |
| | | | | 3 | 570 | 11.0 | 760 | 7.0 | 1140 | 10.0 | 1290 | 17.0 | 800 | 1065 | 1595 | 1805 |
| | | | 15/32 | 2 | 540 | 13.0 | 720 | 7.5 | 1060 | 11.0 | 1200 | 19.0 | 755 | 1010 | 1485 | 1680 |
| | | | | 3 | 600 | 10.0 | 800 | 6.0 | 1200 | 9.0 | 1350 | 15.0 | 840 | 1120 | 1680 | 1890 |
| | 10d | 1-1/2 | 15/32 | 2 | 580 | 25.0 | 770 | 15.0 | 1150 | 21.0 | 1310 | 33.0 | 810 | 1080 | 1610 | 1835 |
| | | | | 3 | 650 | 21.0 | 860 | 12.0 | 1300 | 17.0 | 1470 | 28.0 | 910 | 1205 | 1820 | 2060 |
| | | | 19/32 | 2 | 640 | 21.0 | 850 | 13.0 | 1280 | 18.0 | 1460 | 28.0 | 895 | 1190 | 1790 | 2045 |
| | | | | 3 | 720 | 17.0 | 960 | 10.0 | 1440 | 14.0 | 1640 | 24.0 | 1010 | 1345 | 2015 | 2295 |

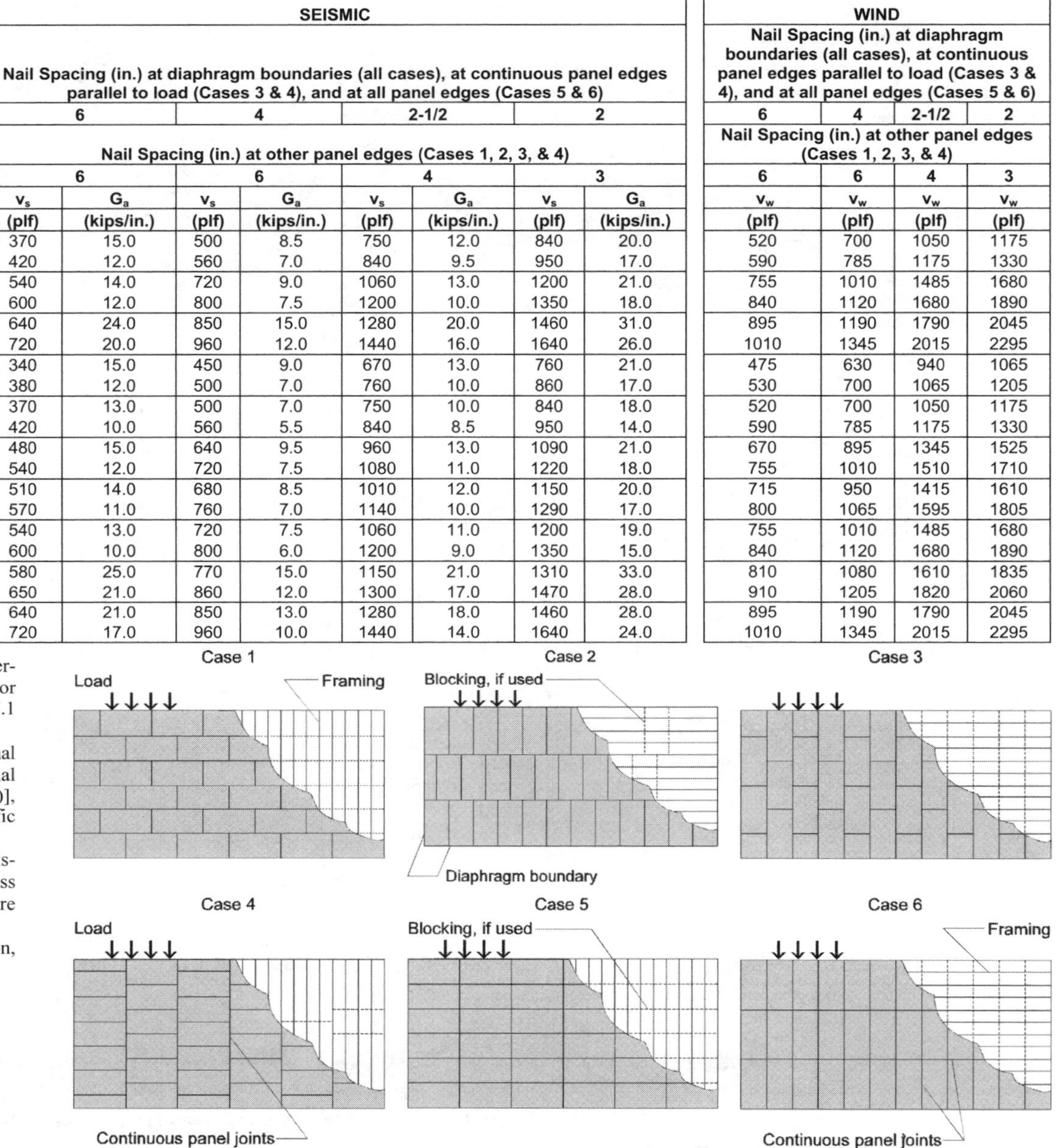


1. Nominal unit shear capacities shall be adjusted in accordance with 4.2.3 to determine ASD allowable unit shear capacity and LRFD factored unit resistance. For general construction requirements see 4.2.6. For specific requirements, see 4.2.7.1 for wood structural panel diaphragms.
2. For framing grades other than Douglas Fir-Larch or Southern Pine, reduced nominal unit shear capacities shall be determined by multiplying the tabulated nominal unit shear capacity by the Specific Gravity Adjustment Factor = [1 – (0.5 – G)], where G = Specific Gravity of the framing lumber from the *NDS*. The Specific Gravity Adjustment Factor shall not be greater than 1.
3. Apparent shear stiffness values, $G_a$, are based on nail slip in framing with moisture content less than or equal to 19% at time of fabrication and panel stiffness values for diaphragms constructed with OSB panels. When plywood panels are used, $G_a$ values shall be determined in accordance with Appendix A.
4. Where moisture content of the framing is greater than 19% at time of fabrication, $G_a$ values shall be multiplied by 0.5.

## Table 4.2B Nominal Unit Shear Capacities for Wood-Frame Diaphragms

### Unblocked Wood Structural Panel Diaphragms (Excluding Plywood for $G_a$)[1,2,3,4]

| Sheathing Grade | Common Nail Size | Minimum Fastener Penetration in Framing (in.) | Minimum Nominal Panel Thickness (in.) | Minimum Nominal Framing Width (in.) | A SEISMIC — Edge Nail Spacing: 6 in. — Case 1 $v_s$ (plf) | Case 1 $G_a$ (kips/in.) | Cases 2,3,4,5,6 $v_s$ (plf) | Cases 2,3,4,5,6 $G_a$ (kips/in.) | B WIND — Edge Nail Spacing: 6 in. — Case 1 $v_w$ (plf) | Cases 2,3,4,5,6 $v_w$ (plf) |
|---|---|---|---|---|---|---|---|---|---|---|
| Structural I | 6d | 1-1/4 | 5/16 | 2 | 330 | 9.0 | 250 | 6.0 | 460 | 350 |
| | | | | 3 | 370 | 7.0 | 280 | 4.5 | 520 | 390 |
| | 8d | 1-3/8 | 3/8 | 2 | 480 | 8.5 | 360 | 6.0 | 670 | 505 |
| | | | | 3 | 530 | 7.5 | 400 | 5.0 | 740 | 560 |
| | 10d | 1-1/2 | 15/32 | 2 | 570 | 14.0 | 430 | 9.5 | 800 | 600 |
| | | | | 3 | 640 | 12.0 | 480 | 8.0 | 895 | 670 |
| Sheathing and Single-Floor | 6d | 1-1/4 | 5/16 | 2 | 300 | 9.0 | 220 | 6.0 | 420 | 310 |
| | | | | 3 | 340 | 7.0 | 250 | 5.0 | 475 | 350 |
| | | | 3/8 | 2 | 330 | 7.5 | 250 | 5.0 | 460 | 350 |
| | | | | 3 | 370 | 6.0 | 280 | 4.0 | 520 | 390 |
| | 8d | 1-3/8 | 3/8 | 2 | 430 | 9.0 | 320 | 6.0 | 600 | 450 |
| | | | | 3 | 480 | 7.5 | 360 | 5.0 | 670 | 505 |
| | | | 7/16 | 2 | 460 | 8.5 | 340 | 5.5 | 645 | 475 |
| | | | | 3 | 510 | 7.0 | 380 | 4.5 | 715 | 530 |
| | | | 15/32 | 2 | 480 | 7.5 | 360 | 5.0 | 670 | 505 |
| | | | | 3 | 530 | 6.5 | 400 | 4.0 | 740 | 560 |
| | 10d | 1-1/2 | 15/32 | 2 | 510 | 15.0 | 380 | 10.0 | 715 | 530 |
| | | | | 3 | 580 | 12.0 | 430 | 8.0 | 810 | 600 |
| | | | 19/32 | 2 | 570 | 13.0 | 430 | 8.5 | 800 | 600 |
| | | | | 3 | 640 | 10.0 | 480 | 7.0 | 895 | 670 |

1. Nominal unit shear capacities shall be adjusted in accordance with 4.2.3 to determine ASD allowable unit shear capacity and LRFD factored unit resistance. For general construction requirements see 4.2.6. For specific requirements, see 4.2.7.1 for wood structural panel diaphragms.
2. For framing grades other than Douglas Fir-Larch or Southern Pine, reduced nominal unit shear capacities shall be determined by multiplying the tabulated nominal unit shear capacity by the Specific Gravity Adjustment Factor = [1 – (0.5 – G)], where G = Specific Gravity of the framing lumber from the *NDS*. The Specific Gravity Adjustment Factor shall not be greater than 1.
3. Apparent shear stiffness values, $G_a$, are based on nail slip in framing with moisture content less than or equal to 19% at time of fabrication and panel stiffness values for diaphragms constructed with OSB panels. When plywood panels are used, $G_a$ values shall be determined in accordance with Appendix A.
4. Where moisture content of the framing is greater than 19% at time of fabrication, $G_a$ values shall be multiplied by 0.5.

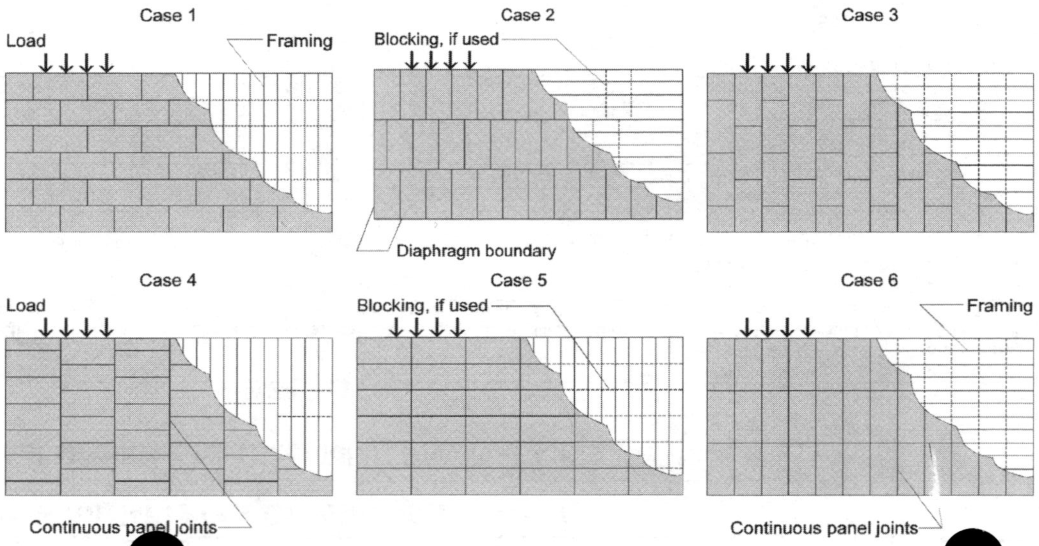

Case 1
Load
Framing
Case 2
Blocking, if used
Case 3
Diaphragm boundary
Case 4
Load
Case 5
Blocking, if used
Case 6
Framing
Continuous panel joints
Continuous panel joints

## Table 4.2C Nominal Unit Shear Capacities for Wood-Frame Diaphragms

### Lumber Diaphragms[1]

| Sheathing Material | Sheathing Nominal Dimensions | Type, Size, and Number of Nails per Board | | A SEISMIC | | B WIND |
|---|---|---|---|---|---|---|
| | | Nailing at Intermediate and End Bearing Supports (Nails/board/support) | Nailing at Boundary Members (Nails/board/end) | $v_s$ (plf) | $G_a$ (kips/in.) | $v_w$ (plf) |
| Horizontal Lumber Sheathing | 1x6 | 2-8d common nails (3-8d box nails) | 3-8d common nails (5-8d box nails) | 100 | 1.5 | 140 |
| | 1x8 | 3-8d common nails (4-8d box nails) | 4-8d common nails (6-8d box nails) | | | |
| | 2x6 | 2-16d common nails (3-16d box nails) | 3-16d common nails (5-16d box nails) | | | |
| | 2x8 | 3-16d common nails (4-16d box nails) | 4-16d common nails (6-16d box nails) | | | |
| Diagonal Lumber Sheathing | 1x6 | 2-8d common nails (3-8d box nails) | 3-8d common nails (5-8d box nails) | 600 | 6 | 840 |
| | 1x8 | 3-8d common nails (4-8d box nails) | 4-8d common nails (6-8d box nails) | | | |
| | 2x6 | 2-16d common nails (3-16d box nails) | 3-16d common nails (5-16d box nails) | | | |
| | 2x8 | 3-16d common nails (4-16d box nails) | 4-16d common nails (6-16d box nails) | | | |
| Double Diagonal Lumber Sheathing | 1x6 | 2-8d common nails (3-8d box nails) | 3-8d common nails (5-8d box nails) | 1200 | 9.5 | 1680 |
| | 1x8 | 3-8d common nails (4-8d box nails) | 4-8d common nails (6-8d box nails) | | | |
| | 2x6 | 2-16d common nails (3-16d box nails) | 3-16d common nails (5-16d box nails) | | | |
| | 2x8 | 3-16d common nails (4-16d box nails) | 4-16d common nails (6-16d box nails) | | | |

1. Nominal unit shear capacities shall be adjusted in accordance with 4.2.3 to determine ASD allowable unit shear capacity and LRFD factored unit resistance. For general construction requirements see 4.2.6. For specific requirements, see 4.2.7.2 for diaphragms diagonally sheathed with a single-layer of lumber, see 4.2.7.3 for diaphragms diagonally sheathed with a double-layer of lumber, and see 4.2.7.4 for diaphragms horizontally sheathed with a single-layer of lumber.

# 4.3 Wood Shear Walls

## 4.3.1 Application Requirements

Wood shear walls are permitted to resist lateral forces provided the deflection of the shear wall, as determined by calculations, tests, or analogies drawn therefrom, does not exceed the maximum permissible deflection limit. Permissible deflection shall be that deflection that permits the shear wall and any attached elements to maintain their structural integrity and continue to support their prescribed loads as determined by the applicable building code or standard.

## 4.3.2 Deflection

Calculations of shear wall deflection shall account for bending and shear deflections, fastener deformation, anchorage slip, and other contributing sources of deflection.

The shear wall deflection, $\delta_{sw}$, is permitted to be calculated by use of the following equation:

$$\delta_{sw} = \frac{8vh^3}{EAb} + \frac{vh}{1000G_a} + \frac{h\Delta_a}{b} \quad (4.3\text{-}1)$$

**where:**

$b$ = shear wall length, ft

$\Delta_a$ = total vertical elongation of wall anchorage system (including fastener slip, device elongation, rod elongation, etc.) at the induced unit shear in the shear wall, in.

$E$ = modulus of elasticity of end posts, psi

$A$ = area of end post cross-section, in.$^2$

$G_a$ = apparent shear wall shear stiffness from nail slip and panel shear deformation, kips/in. (from Column A, Tables 4.3A, 4.3B, or 4.3C)

$h$ = shear wall height, ft

$v$ = induced unit shear, lbs/ft

$\delta_{sw}$ = maximum shear wall deflection determined by elastic analysis, in.

Alternatively, for wood structural panel shear walls, deflection is permitted to be calculated using a rational analysis where apparent shear stiffness accounts for panel shear deformation and non-linear nail slip in the sheathing to framing connection.

4.3.2.1 Deflection of Perforated Shear Walls: The deflection of a perforated shear wall shall be calculated in accordance with 4.3.2, where $v$ in Equation 4.3-1 is equal to $v_{max}$ obtained in Equation 4.3-6, and $b$ is taken as $\Sigma L_i$.

## 4.3.3 Unit Shear Capacities

The ASD allowable unit shear capacity shall be determined by dividing the tabulated nominal unit shear capacity, modified by applicable footnotes, by the ASD reduction factor of 2.0. No further increases shall be permitted. The LRFD factored unit resistance shall be determined by multiplying the nominal unit shear capacity by a resistance factor, $\phi_D$, of 0.80.

4.3.3.1 Tabulated Nominal Unit Shear Capacities: Tabulated nominal unit shear capacities for seismic design are provided in Column A of Tables 4.3A, 4.3B, and 4.3C; and for wind design in Column B of Tables 4.3A, 4.3B, and 4.3C.

4.3.3.2 Summing Shear Capacities: For shear walls sheathed with the same construction and materials on opposite sides of the same wall, the combined nominal unit shear capacity, $v_{sc}$ or $v_{wc}$, shall be permitted to be taken as twice the nominal unit shear capacity for an equivalent shear wall sheathed on one side.

4.3.3.2.1 For seismic design of shear walls sheathed with the same construction and materials on opposite sides of a shear wall, the shear wall deflection shall be calculated using the combined apparent shear wall shear stiffness, $G_{ac}$, and the combined nominal unit shear capacity, $v_{sc}$, using the following equations:

$$G_{ac} = G_{a1} + G_{a2} \quad (4.3\text{-}2)$$

$$v_{sc} = K_{min} G_{ac} \quad (4.3\text{-}3)$$

**where:**

$G_{ac}$ = combined apparent shear wall shear stiffness of two-sided shear wall, kips/in.

$G_{a1}$ = apparent shear wall shear stiffness for side 1, kips/in. (from Column A, Tables 4.3A, 4.3B, or 4.3C)

$G_{a2}$ = apparent shear wall shear stiffness for side 2, kips/in. (from Column A, Tables 4.3A, 4.3B, or 4.3C)

$K_{min}$ = minimum ratio of $v_{s1}/G_{a1}$ or $v_{s2}/G_{a2}$

$v_{s1}$ = nominal unit shear capacity for side 1, lbs/ft (from Column A, Tables 4.3A, 4.3B, or 4.3C)

$v_{s2}$ = nominal unit shear capacity for side 2, lbs/ft (from Column A, Tables 4.3A, 4.3B, or 4.3C)

$v_{sc}$ = Combined nominal unit shear capacity of two-sided shear wall for seismic design, lbs/ft

4.3.3.2.2 Nominal unit shear capacities for shear walls sheathed with dissimilar materials on the same side of the wall are not cumulative. For shear walls sheathed with dissimilar materials on opposite sides, the combined nominal unit shear capacity, $v_{sc}$ or $v_{wc}$, shall be either two times the smaller nominal unit shear capacity or the larger nominal unit shear capacity, whichever is greater.

**Exception:** For wind design, the combined nominal unit shear capacity, $v_{wc}$, of shear walls sheathed with a combination of wood structural panels, hardboard panel siding, or structural fiberboard on one side and gypsum wallboard on the opposite side shall equal the sum of the sheathing capacities of each side separately.

4.3.3.3 Summing Shear Wall Lines: The nominal shear capacity for shear walls in a line, utilizing shear walls sheathed with the same materials and construction, shall be permitted to be combined if the induced shear load is distributed so as to provide the same deflection, $\delta_{SW}$, in each shear wall. Summing nominal unit shear capacities of dissimilar materials applied to the same wall line is not allowed.

4.3.3.4 Shear Capacity of Perforated Shear Walls: The nominal shear capacity of a perforated shear wall shall be taken as the nominal unit shear capacity multiplied by the sum of the shear wall segment lengths, $\Sigma L_i$, and the appropriate shear capacity adjustment factor, $C_o$, from Table 4.3.3.4.

### Table 4.3.3.4 Shear Capacity Adjustment Factor, $C_o$

| Wall Height, h | Maximum Opening Height[1] | | | | |
|---|---|---|---|---|---|
| | h/3 | h/2 | 2h/3 | 5h/6 | h |
| 8' Wall | 2' - 8" | 4' - 0" | 5' - 4" | 6' - 8" | 8' - 0" |
| 10' Wall | 3' - 4" | 5' – 0" | 6' - 8" | 8' - 4" | 10' - 0" |
| **Percent Full-Height Sheathing[2]** | **Effective Shear Capacity Ratio** | | | | |
| 10% | 1.00 | 0.69 | 0.53 | 0.43 | 0.36 |
| 20% | 1.00 | 0.71 | 0.56 | 0.45 | 0.38 |
| 30% | 1.00 | 0.74 | 0.59 | 0.49 | 0.42 |
| 40% | 1.00 | 0.77 | 0.63 | 0.53 | 0.45 |
| 50% | 1.00 | 0.80 | 0.67 | 0.57 | 0.50 |
| 60% | 1.00 | 0.83 | 0.71 | 0.63 | 0.56 |
| 70% | 1.00 | 0.87 | 0.77 | 0.69 | 0.63 |
| 80% | 1.00 | 0.91 | 0.83 | 0.77 | 0.71 |
| 90% | 1.00 | 0.95 | 0.91 | 0.87 | 0.83 |
| 100% | 1.00 | 1.00 | 1.00 | 1.00 | 1.00 |

1. The maximum opening height shall be taken as the maximum opening clear height in a perforated shear wall. Where areas above and/or below an opening remain unsheathed, the height of each opening shall be defined as the clear height of the opening plus the unsheathed areas.
2. The sum of the lengths of the perforated shear wall segments divided by the total length of the perforated shear wall.

## 4.3.4 Shear Wall Aspect Ratios

Size and shape of shear walls shall be limited to the aspect ratios in Table 4.3.4.

### Table 4.3.4 Maximum Shear Wall Aspect Ratios

| Shear Wall Sheathing Type | Maximum $h/b_s$ Ratio |
|---|---|
| Wood structural panels, all edges nailed | 3-½:1[1] |
| Particleboard, all edges nailed | 2:1 |
| Diagonal sheathing, conventional | 2:1 |
| Gypsum wallboard | 2:1[2] |
| Portland cement plaster | 2:1[2] |
| Fiberboard | 1-½:1 |

1. For design to resist seismic forces, the shear wall aspect ratio shall not exceed 2:1 unless the nominal unit shear capacity is multiplied by $2b_s/h$.
2. Walls having aspect ratios exceeding 1-½:1 shall be blocked.

4.3.4.1 Aspect Ratio of Perforated Shear Wall Segments: The aspect ratio limitations of 4.3.4 shall apply to perforated shear wall segments within a perforated shear wall. For design to resist seismic forces, the nominal shear capacity of the perforated shear wall shall be multiplied by $2b_s/h$ when the aspect ratio of the narrowest perforated shear wall segment included in the sum of shear wall segment lengths, $\Sigma L_i$, is greater than 2:1, but does not exceed 3-½:1. Portions of walls in excess of 3-½:1 shall not be counted in the sum of shear wall segments.

## 4.3.5 Shear Wall Types

Where individual full-height wall segments are designed as shear walls, provisions of 4.3.5.1 shall apply. For shear walls with openings, where framing and connections around the openings are designed for force transfer around the openings the provisions of 4.3.5.2 shall apply. For shear walls with openings, where framing and connections around the opening are not designed for force transfer around the openings (perforated shear walls) the provisions of 4.3.5.3 shall apply or individual full-height wall segments shall be designed per 4.3.5.1.

4.3.5.1 Segmented Shear Walls: Where full-height wall segments are designed as shear walls, aspect ratio limitations of 4.3.4 shall apply to each full-height wall segment.

4.3.5.2 Force Transfer Around Openings: Where shear walls with openings are designed for force transfer around the openings, the aspect ratio limitations of 4.3.4 shall apply to the overall shear wall including openings and to each wall pier at the sides of an opening. The height of a wall pier shall be defined as the clear height of the pier at the side of an opening. The length of a wall pier shall be defined as the sheathed length of the pier. Design for force transfer shall be based on a rational analysis. The length of a wall pier shall not be less than 2'.

4.3.5.3 Perforated Shear Walls: Where wood structural panel shear walls with openings are not designed for force transfer around the opening, they shall be designed as perforated shear walls. The following limitations shall apply:

1. A perforated shear wall segment shall be located at each end of a perforated shear wall. Openings shall be permitted to occur beyond the ends of the perforated shear wall. However, the length of such openings shall not be included in the length of the perforated shear wall.
2. The nominal unit shear capacity for a single-sided wall shall not exceed 980 plf for seismic or 1,370 plf for wind as given in Table 4.3A. The nominal unit shear capacity for a double-sided wall shall not exceed 2,000 plf for wind.
3. Where out-of-plane offsets occur, portions of the wall on each side of the offset shall be considered as separate perforated shear walls.
4. Collectors for shear transfer shall be provided through the full length of the perforated shear wall.
5. A perforated shear wall shall have uniform top-of-wall and bottom-of-wall elevations. Perforated shear walls not having uniform elevations shall be designed by other methods.
6. Perforated shear wall height, h, shall not exceed 20'.

## 4.3.6 Construction Requirements

4.3.6.1 Framing Requirements: All framing used for shear wall construction shall be 2" nominal or larger members. Shear wall boundary elements, such as end posts, shall be provided to transmit the design tension and compression forces. Shear wall sheathing shall not be used to splice boundary elements. End posts (studs or columns) shall be framed to provide full end bearing.

4.3.6.1.1 Tension and Compression Chords: Tension force, T, and a compression force, C, resulting from shear wall overturning forces at each story level shall be calculated in accordance with the following:

$$T = C = vh \qquad (4.3\text{-}4)$$

**where:**

C = compression force, lbs

h = shear wall height, ft

T = tension force, lbs

$v$ = induced unit shear, lbs/ft

4.3.6.1.2 Tension and Compression Chords of Perforated Shear Walls: Each end of each perforated shear wall shall be designed for a tension force, T, and a compression force, C. Each end of each perforated shear wall segment shall be designed for a compression force, C, in each segment. For perforated shear walls, the values for T and C resulting from shear wall overturning at each story level shall be calculated in accordance with the following:

$$T = C = \frac{Vh}{C_o \sum L_i} \qquad (4.3\text{-}5)$$

**where:**

$C_o$ = shear capacity adjustment factor from Table 4.3.3.4

V = induced shear force in perforated shear wall, lbs

$\Sigma L_i$ = sum of perforated shear wall segment lengths, ft

4.3.6.2 Sheathing: Shear walls shall be sheathed with approved materials. Details on sheathing types and thicknesses for commonly used shear wall assemblies are provided in 4.3.7 and Tables 4.3A, 4.3B, and 4.3C.

4.3.6.3 Fasteners: Sheathing shall be attached to framing using approved fasteners. Nails or other approved sheathing fasteners shall be driven with the head of the fastener flush with the surface of the sheathing. Details on type, size, and spacing of mechanical fasteners in commonly used shear wall assemblies are provided in 4.3.7 and Tables 4.3A, 4.3B, and 4.3C.

4.3.6.3.1 Adhesives: Adhesive attachment of shear wall sheathing shall not be used alone, or in combination with mechanical fasteners.

**Exception:** Approved adhesive attachment systems shall be permitted for wind and seismic design in Seismic Design Categories A, B, and C where R = 1.5 and $\Omega_0$ = 2.5, unless other values are approved.

4.3.6.4 Shear Wall Anchorage and Load Path: Design of shear wall anchorage and load path shall conform to the requirements of this section, or shall be calculated using principles of mechanics.

4.3.6.4.1 Anchorage for In-plane Shear: Connections shall be provided to transfer the induced unit shear force, $v$, into and out of each shear wall.

4.3.6.4.1.1 In-plane Shear Anchorage for Perforated Shear Walls: The maximum induced unit shear force, $v_{max}$, transmitted into the top of a perforated shear wall, out of the base of the perforated shear wall at full height sheathing, and into collectors connecting shear wall segments, shall be calculated in accordance with the following:

$$v_{max} = \frac{V}{C_o \sum L_i} \qquad (4.3\text{-}6)$$

4.3.6.4.2 Uplift Anchorage at Shear Wall Ends: Where the dead load stabilizing moment is not sufficient to prevent uplift due to overturning moments on the wall (from 4.3.6.1.1), an anchoring device shall be provided at the end of each shear wall.

4.3.6.4.2.1 Uplift Anchorage for Perforated Shear Walls: In addition to the requirements of 4.3.6.4.2, perforated shear wall bottom plates at full height sheathing shall be anchored for a uniform uplift force, t, equal to the unit shear force, $v_{max}$, determined in 4.3.6.4.1.1, or calculated by rational analysis.

4.3.6.4.3 Anchor Bolts: Foundation anchor bolts shall have a steel plate washer under each nut not less than 2-½" x 2-½" x ¼". The plate washer shall extend to within ½" of the edge of the bottom plate on the sheathed side.

4.3.6.4.4 Load Path: A load path to the foundation shall be provided for uplift, shear, and compression forces. Elements resisting shear wall forces contributed by multiple stories shall be designed for the sum of forces contributed by each story.

## 4.3.7 Shear Wall Systems

4.3.7.1 Wood Structural Panel Shear Walls: Shear walls sheathed with wood structural panel sheathing shall be permitted to be used to resist seismic and wind forces. The size and spacing of fasteners at shear wall boundaries, panel edges, and intermediate supports shall be as provided in Table 4.3A. The shear wall shall be constructed as follows:

1. Panels shall not be less than 4' x 8', except at boundaries and changes in framing. Framing members or blocking shall be provided at the edges of all panels.
2. Nails located at least 3/8" from edges and ends of panels. Maximum nail spacing of 6" on center at panel edges. Maximum nail spacing of 6" on center along intermediate framing members for 3/8" and 7/16" panels installed on studs spaced 24" on center. Maximum nail spacing along intermediate framing of 12" for thicker panels or closer stud spacings.
3. 2" nominal or wider framing thickness at adjoining panel edges except that 3" nominal or wider framing thickness and staggered nailing are required where:
   a. Nails are spaced 2" on center or less at adjoining panel edges, or
   b. 10d nails having penetration into framing of more than 1-½" are spaced 3" on center, or less at adjoining panel edges, or
   c. Required nominal unit shear capacity exceeds 700 plf in seismic Design Category D, E, or F.
4. Maximum stud spacing of 24" on center.
5. Wood structural panels shall conform to the requirements for its type in DOC PS 1 or PS 2.

4.3.7.2 Particleboard Shear Walls: Shear walls sheathed with particleboard sheathing shall be permitted to be used to resist wind forces and seismic forces in Seismic Design Categories A, B, and C. The size and spacing of fasteners at shear wall boundaries, panel edges, and intermediate supports shall be as provided in Table 4.3A. The shear wall shall be constructed as follows:

1. Panels shall not be less than 4' x 8', except at boundaries and changes in framing. Framing members or blocking shall be provided at the edges of all panels.
2. Nails located at least 3/8" from edges and ends of panels. Maximum nail spacing of 6" on center along intermediate framing members for 3/8" panels installed on studs spaced 24" on center. Maximum nail spacing along intermediate framing of 12" on center for thicker panels or closer stud spacings.
3. 2" nominal or wider framing thickness at adjoining panel edges except that 3" nominal or wider framing thickness and staggered nailing are required where:
   a. Nails are spaced 2" on center or less at adjoining panel edges, or
   b. 10d nails having penetration into framing of more than 1-½" are spaced 3" on center, or less at adjoining panel edges.
4. Maximum stud spacing of 24" on center.
5. Particleboard shall conform to ANSI A208.1.

4.3.7.3 Fiberboard Shear Walls: Shear walls sheathed with fiberboard sheathing shall be permitted to be used to resist wind forces and seismic forces in Seismic Design Categories A, B, and C. The size and spacing of fasteners at shear wall boundaries, panel edges, and intermediate supports shall be as provided in Table 4.3A. The shear wall shall be constructed as follows:

1. Panels shall not be less than 4' x 8', except at boundaries and changes in framing. Framing members or blocking shall be provided at the edges of all panels.
2. Nails located at least 3/8" from edges and ends of panels. Maximum nail spacing of 6" on center along intermediate framing members.
3. 2" nominal or wider framing at adjoining panel edges.
4. Maximum stud spacing of 16" on center.
5. Minimum length of galvanized roofing nails is 1-½" for ½" thick sheathing and 1-¾" for 25/32" thick sheathing.
6. Fiberboard sheathing shall conform to either AHA 194.1 or ASTM C 208.

4.3.7.4 Gypsum Wallboard, Gypsum Veneer Base, Water-Resistant Backing Board, Gypsum Sheathing, Gypsum Lath and Plaster, or Portland Cement Plaster Shear Walls: Shear walls sheathed with gypsum wallboard, gypsum veneer base, water-resistant backing board, gypsum sheathing, gypsum lath and plaster, or portland cement plaster shall be permitted to be used to resist wind forces and seismic forces in Seismic Design Categories A through D. End joints of adjacent courses of gypsum wallboard or sheathing shall not occur over the same stud. The size and spacing of fasteners at shear wall boundaries, panel edges, and intermediate supports shall be as provided in Table 4.3B. Nails shall be spaced not less than 3/8" from edges and ends of panels. Wood framing shall be 2" nominal or wider.

4.3.7.4.1 Gypsum Wallboard, Gypsum Veneer Base, Water-Resistant Gypsum Backing Board: Gypsum wallboard, gypsum veneer base, or water-resistant gypsum backing board shall be applied parallel or perpendicular to studs. Gypsum wallboard shall conform to ASTM C 36 and shall be installed in accordance with ASTM C 840. Gypsum veneer base shall conform to ASTM C 588 and shall be installed in accordance with ASTM C 844. Water-resistant backing board shall conform to ASTM C 630 and shall be installed in accordance with ASTM C 840.

4.3.7.4.2 Gypsum Sheathing: Four-foot-wide pieces of gypsum sheathing shall be applied parallel or perpendicular to studs. Two-foot-wide pieces of gypsum sheathing shall be applied perpendicular to the studs. Gypsum sheathing shall conform to ASTM C 79 and shall be installed in accordance with ASTM C 1280.

4.3.7.4.3 Gypsum Lath and Plaster: Gypsum lath shall be applied perpendicular to the studs. Gypsum lath shall conform to ASTM C 37 and shall be installed in accordance with ASTM C 841. Gypsum plaster shall conform to the requirements of ASTM C 28.

4.3.7.4.4 Expanded Metal or Woven Wire Lath and Portland Cement: Expanded metal or woven wire lath and portland cement shall conform to ASTM C 847, ASTM C 1032, and ASTM C 150 and shall be installed in accordance with ASTM C 926 and ASTM C 1063. Metal lath and lath attachments shall be of corrosion-resistant material.

4.3.7.5 Shear Walls Diagonally Sheathed with Single-Layer of Lumber: Single diagonally sheathed lumber shear walls are permitted to be used to resist wind forces and seismic forces in Seismic Design Categories A, B, C, and D. Single diagonally sheathed lumber shear walls shall be constructed of minimum 1" thick nominal sheathing boards laid at an angle of approximately 45° to the supports. End joints in adjacent boards shall be separated by at least one stud space and there shall be at least two boards between joints on the same support. Nailing of diagonally sheathed lumber shear walls shall be in accordance with Table 4.3C.

4.3.7.6 Shear Walls Diagonally Sheathed with Double-Layer of Lumber: Double diagonally sheathed lumber shear walls are permitted to be used to resist wind forces and seismic forces in Seismic Design Categories A, B, C, and D. Double diagonally sheathed lumber shear walls shall be constructed of two layers of 1" thick nominal diagonal sheathing boards laid perpendicular to each other on the same face of the supporting members. Nailing of diagonally sheathed lumber shear walls shall be in accordance with Table 4.3C.

4.3.7.7 Shear Walls Horizontally Sheathed with Single-Layer of Lumber: Horizontally sheathed lumber shear walls are permitted to be used to resist wind forces and seismic forces in Seismic Design Categories A, B, and C. Horizontally sheathed lumber shear walls shall be constructed of minimum 1" thick nominal sheathing boards applied perpendicular to the supports. End joints in adjacent boards shall be separated by at least one stud space and there shall be at least two boards between joints on the same support. Nailing of horizontally sheathed lumber shear walls shall be in accordance with Table 4.3C.

4.3.7.8 Shear Walls Sheathed with Vertical Board Siding: Vertical board siding shear walls are permitted to be used to resist wind forces and seismic forces in Seismic Design Categories A, B, and C. Vertical board siding shear walls shall be constructed of minimum 1" thick nominal sheathing boards applied directly to studs and blocking. Nailing of vertical board siding shear walls shall be in accordance with Table 4.3C.

# Table 4.3A Nominal Unit Shear Capacities for Wood-Frame Shear Walls[1,3]

## Wood-based Panels (Excluding Plywood for $G_a$)[4]

| Sheathing Material | Minimum Nominal Panel Thickness (in.) | Minimum Fastener Penetration in Framing (in.) | Fastener Type & Size | A SEISMIC | | | | | | | | B WIND | | | |
|---|---|---|---|---|---|---|---|---|---|---|---|---|---|---|---|
| | | | | Panel Edge Fastener Spacing (in.) | | | | | | | | Panel Edge Fastener Spacing (in.) | | | |
| | | | | 6 | | 4 | | 3 | | 2 | | 6 | 4 | 3 | 2 |
| | | | | $v_s$ | $G_a$ | $v_s$ | $G_a$ | $v_s$ | $G_a$ | $v_s$ | $G_a$ | $v_w$ | $v_w$ | $v_w$ | $v_w$ |
| | | | | (plf) | (kips/in.) | (plf) | (kips/in.) | (plf) | (kips/in.) | (plf) | (kips/in.) | (plf) | (plf) | (plf) | (plf) |
| Wood Structural Panels - Structural I[4,5] | | | Nail (common or galvanized box) | | | | | | | | | | | | |
| | 5/16 | 1-1/4 | 6d | 400 | 13.0 | 600 | 18.0 | 780 | 23.0 | 1020 | 35.0 | 560 | 840 | 1090 | 1430 |
| | 3/8[2] | 1-3/8 | 8d | 460 | 19.0 | 720 | 24.0 | 920 | 30.0 | 1220 | 43.0 | 645 | 1010 | 1290 | 1710 |
| | 7/16[2] | | | 510 | 16.0 | 790 | 21.0 | 1010 | 27.0 | 1340 | 40.0 | 715 | 1105 | 1415 | 1875 |
| | 15/32 | | | 560 | 14.0 | 860 | 18.0 | 1100 | 24.0 | 1460 | 37.0 | 785 | 1205 | 1540 | 2045 |
| | 15/32 | 1-1/2 | 10d | 680 | 22.0 | 1020 | 29.0 | 1330 | 36.0 | 1740 | 51.0 | 950 | 1430 | 1860 | 2435 |
| Wood Structural Panels – Sheathing[4,5] | 5/16 | 1-1/4 | 6d | 360 | 13.0 | 540 | 18.0 | 700 | 24.0 | 900 | 37.0 | 505 | 755 | 980 | 1260 |
| | 3/8 | | | 400 | 11.0 | 600 | 15.0 | 780 | 20.0 | 1020 | 32.0 | 560 | 840 | 1090 | 1430 |
| | 3/8[2] | 1-3/8 | 8d | 440 | 17.0 | 640 | 25.0 | 820 | 31.0 | 1060 | 45.0 | 615 | 895 | 1150 | 1485 |
| | 7/16[2] | | | 480 | 15.0 | 700 | 22.0 | 900 | 28.0 | 1170 | 42.0 | 670 | 980 | 1260 | 1640 |
| | 15/32 | | | 520 | 13.0 | 760 | 19.0 | 980 | 25.0 | 1280 | 39.0 | 730 | 1065 | 1370 | 1790 |
| | 15/32 | 1-1/2 | 10d | 620 | 22.0 | 920 | 30.0 | 1200 | 37.0 | 1540 | 52.0 | 870 | 1290 | 1680 | 2155 |
| | 19/32 | | | 680 | 19.0 | 1020 | 26.0 | 1330 | 33.0 | 1740 | 48.0 | 950 | 1430 | 1860 | 2435 |
| Plywood Siding | | | Nail (galvanized casing) | | | | | | | | | | | | |
| | 5/16 | 1-1/4 | 6d | 280 | 13.0 | 420 | 16.0 | 550 | 17.0 | 720 | 21.0 | 390 | 590 | 770 | 1010 |
| | 3/8 | 1-3/8 | 8d | 320 | 16.0 | 480 | 18.0 | 620 | 20.0 | 820 | 22.0 | 450 | 670 | 870 | 1150 |
| Particleboard Sheathing - (M-S "Exterior Glue" and M-2 "Exterior Glue") | | | Nail (common or galvanized box) | | | | | | | | | | | | |
| | 3/8 | | 6d | 240 | 15.0 | 360 | 17.0 | 460 | 19.0 | 600 | 22.0 | 335 | 505 | 645 | 840 |
| | 3/8 | | 8d | 260 | 18.0 | 380 | 20.0 | 480 | 21.0 | 630 | 23.0 | 365 | 530 | 670 | 880 |
| | 1/2 | | | 280 | 18.0 | 420 | 20.0 | 540 | 22.0 | 700 | 24.0 | 390 | 590 | 755 | 980 |
| | 1/2 | | 10d | 370 | 21.0 | 550 | 23.0 | 720 | 24.0 | 920 | 25.0 | 520 | 770 | 1010 | 1290 |
| | 5/8 | | | 400 | 21.0 | 610 | 23.0 | 790 | 24.0 | 1040 | 26.0 | 560 | 855 | 1105 | 1455 |
| Fiberboard Sheathing - Structural | | | Nail (common or galvanized roofing) | | | | | | | | | | | | |
| | 1/2 | | 8d common or 11 ga. galv. roofing nail (0.120" x 1-1/2" long x 7/16" head) | | | 340 | 4.0 | 460 | 5.0 | 520 | 5.5 | | 475 | 645 | 730 |
| | 25/32 | | 8d common or 11 ga. galv. roofing nail (0.120" x 1-3/4" long x 7/16" head) | | | 360 | 4.0 | 480 | 5.0 | 540 | 5.5 | | 505 | 670 | 755 |

1. Nominal unit shear capacities shall be adjusted in accordance with 4.3.3 to determine ASD allowable unit shear capacity and LRFD factored unit resistance. For general construction requirements see 4.3.6. For specific requirements, see 4.3.7.1 for wood structural panel shear walls, 4.3.7.2 for particleboard shear walls, and 4.3.7.3 for fiberboard shear walls.
2. Shears are permitted to be increased to values shown for 15/32 inch sheathing with same nailing provided (a) studs are spaced a maximum of 16 inches on center, or (b) panels are applied with long dimension across studs.
3. For framing grades other than Douglas Fir-Larch or Southern Pine, reduced nominal unit shear capacities shall be determined by multiplying the tabulated nominal unit shear capacity by the Specific Gravity Adjustment Factor = [1 – (0.5 – G)], where G = Specific Gravity of the framing lumber from the *NDS*. The Specific Gravity Adjustment Factor shall not be greater than 1.
4. Apparent shear stiffness values, $G_a$, are based on nail slip in framing with moisture content less than or equal to 19% at time of fabrication and panel stiffness values for shear walls constructed with OSB panels. When plywood panels are used, $G_a$ values shall be determined in accordance with Appendix A.
5. Where moisture content of the framing is greater than 19% at time of fabrication, $G_a$ values shall be multiplied by 0.5.

## Table 4.3B Nominal Unit Shear Capacities for Wood-Frame Shear Walls[1]

### Gypsum and Portland Cement Plaster

| Sheathing Material | Material Thickness | Fastener Type & Size[2] | Max. Fastener Edge Spacing[3] | Max. Stud Spacing | | A SEISMIC $v_s$ (plf) | A SEISMIC $G_a$ (kips/in.) | B WIND $v_w$ (plf) |
|---|---|---|---|---|---|---|---|---|
| Gypsum wallboard, gypsum veneer base, or water-resistant gypsum backing board | 1/2" | 5d cooler (0.086" x 1-5/8" long, 15/64" head) or wallboard nail (0.086" x 1-5/8" long, 9/32" head) or 0.120" nail x 1-1/2" long, min. 3/8" head | 7" | 24" | unblocked | 150 | 4.0 | 150 |
| | | | 4" | 24" | unblocked | 220 | 6.0 | 220 |
| | | | 7" | 16" | unblocked | 200 | 5.5 | 200 |
| | | | 4" | 16" | unblocked | 250 | 6.5 | 250 |
| | | | 7" | 16" | blocked | 250 | 6.5 | 250 |
| | | | 4" | 16" | blocked | 300 | 7.5 | 300 |
| | | No. 6 Type S or W drywall screws 1-1/4" long | 8/12" | 16" | unblocked | 120 | 3.5 | 120 |
| | | | 4/16" | 16" | blocked | 320 | 8.0 | 320 |
| | | | 4/12" | 24" | blocked | 310 | 8.0 | 310 |
| | | | 8/12" | 16" | blocked | 140 | 4.0 | 140 |
| | | | 6/12" | 16" | blocked | 180 | 5.0 | 180 |
| | 5/8" | 6d cooler (0.092" x 1-7/8" long, 1/4" head) or wallboard nail (0.0915" x 1-7/8" long, 19/64" head) or 0.120" nail x 1-3/4" long, min. 3/8" head | 7" | 24" | unblocked | 230 | 6.0 | 230 |
| | | | 4" | 24" | unblocked | 290 | 7.5 | 290 |
| | | | 7" | 16" | blocked | 290 | 7.5 | 290 |
| | | | 4" | 16" | blocked | 350 | 8.5 | 350 |
| | | No. 6 Type S or W drywall screws 1-1/4" long | 8/12" | 16" | unblocked | 140 | 4.0 | 140 |
| | | | 8/12" | 16" | blocked | 180 | 5.0 | 180 |
| | 5/8" (Two-Ply) | Base ply–6d cooler (0.092" x 1-7/8" long, 1/4" head) or wallboard nail (0.0915" x 1-7/8" long, 19/64" head) or 0.120" nail x 1-3/4" long, min. 3/8" head<br>Face ply–8d cooler (0.113" x 2-3/8" long, 0.281" head) or wallboard nail (0.113" x 2-3/8" long, 3/8" head) or 0.120" nail x 2-3/8" long, min. 3/8" head | Base: 9"<br>Face: 7" | 16" | blocked | 500 | 11.0 | 500 |
| Gypsum sheathing | 1/2" x 2' x 8' | 0.120" nail x 1-3/4" long, 7/16" head, diamond-point, galvanized | 4" | 16" | unblocked | 150 | 4.0 | 150 |
| | 1/2" x 4' | | 4" | 24" | blocked | 350 | 8.5 | 350 |
| | | | 7" | 16" | unblocked | 200 | 5.5 | 200 |
| | 5/8" x 4' | 6d galvanized cooler (0.092" x 1-7/8" long, 1/4" head) or wallboard nail (0.0915" x 1-7/8" long, 19/64" head) or 0.120" nail x 1-3/4" long, min. 3/8" head | 4/7" | 16" | blocked | 400 | 9.5 | 400 |
| Gypsum lath, plain or perforated | 3/8" lath and 1/2" plaster | 0.092" x 1-1/8" long, 19/64" head, gypsum wallboard blued nail or 0.120" nail x 1-1/4" long, min 3/8" head | 5" | 16" | unblocked | 200 | 5.5 | 200 |
| Expanded metal or woven wire lath and Portland cement plaster | 7/8" | 0.120" nail x 1-1/2" long, 7/16" head | 6" | 16" | unblocked | 360 | 9.0 | 360 |

1. Nominal unit shear capacities shall be adjusted in accordance with 4.3.3 to determine ASD allowable unit shear capacity and LRFD factored unit resistance. For general construction requirements see 4.3.6. For specific requirements, see 4.3.7.4.
2. Type S or W drywall screws shall conform to requirements of ASTM C 1002.
3. Where two numbers are given for maximum fastener edge spacing, the first number denotes fastener spacing at the edges and the second number denotes fastener spacing in the field.

## Table 4.3C Nominal Unit Shear Capacities for Wood-Frame Shear Walls[1]

### Lumber Shear Walls

| Sheathing Material | Sheathing Nominal Dimensions | Type, Size, and Number of Nails per Board | | A SEISMIC | | B WIND |
|---|---|---|---|---|---|---|
| | | Nailing at Intermediate Studs (nails/board/support) | Nailing at Shear Wall Boundary Members (nails/board/end) | $v_s$ (plf) | $G_a$ (kips/in.) | $v_w$ (plf) |
| Horizontal Lumber Sheathing | 1x6 & smaller | 2-8d common nails (3-8d box nails) | 3-8d common nails (5-8d box nails) | 100 | 1.5 | 140 |
| | 1x8 & larger | 3-8d common nails (4-8d box nails) | 4-8d common nails (6-8d box nails) | | | |
| Diagonal Lumber Sheathing | 1x6 & smaller | 2-8d common nails (3-8d box nails) | 3-8d common nails (5-8d box nails) | 600 | 6.0 | 840 |
| | 1x8 & larger | 3-8d common nails (4-8d box nails) | 4-8d common nails (6-8d box nails) | | | |
| Double Diagonal Lumber Sheathing | 1x6 & smaller | 2-8d common nails (3-8d box nails) | 3-8d common nails (5-8d box nails) | 1200 | 10.0 | 1680 |
| | 1x8 & larger | 3-8d common nails (4-8d box nails) | 4-8d common nails (6-8d box nails) | | | |
| Vertical Lumber Siding | 1x6 & smaller | 2-8d common nails (3-8d box nails) | 3-8d common nails (5-8d box nails) | 90 | 1.0 | 125 |
| | 1x8 & larger | 3-8d common nails (4-8d box nails) | 4-8d common nails (6-8d box nails) | | | |

1. Nominal unit shear capacities shall be adjusted in accordance with 4.3.3 to determine ASD allowable unit shear capacity and LRFD factored unit resistance. For general construction requirements see 4.3.6. For specific requirements, see 4.3.7.5 through 4.3.7.8.

# APPENDIX A

# Table A.4.2A Nominal Unit Shear Capacities for Wood-Frame Plywood Diaphragms

## Blocked Wood Structural Panel Diaphragms[1,2,3,4]

| Sheathing Grade | Common Nail Size | Minimum Fastener Penetration in Framing (in.) | Minimum Nominal Panel Thickness (in.) | Minimum Nominal Framing Width (in.) | A SEISMIC | | | | | | | | B WIND | | | |
|---|---|---|---|---|---|---|---|---|---|---|---|---|---|---|---|---|
| | | | | | Nail Spacing (in.) at diaphragm boundaries (all cases), at continuous panel edges parallel to load (Cases 3 & 4), and at all panel edges (Cases 5 & 6) | | | | | | | | Nail Spacing (in.) at diaphragm boundaries (all cases), at continuous panel edges parallel to load (Cases 3 & 4), and at all panel edges (Cases 5 & 6) | | | |
| | | | | | 6 | | 4 | | 2-1/2 | | 2 | | 6 | 4 | 2-1/2 | 2 |
| | | | | | Nail Spacing (in.) at other panel edges (Cases 1, 2, 3, & 4) | | | | | | | | Nail Spacing (in.) at other panel edges (Cases 1, 2, 3, & 4) | | | |
| | | | | | 6 | | 6 | | 4 | | 3 | | 6 | 6 | 4 | 3 |
| | | | | | $v_s$ | $G_a$ | $v_s$ | $G_a$ | $v_s$ | $G_a$ | $v_s$ | $G_a$ | $v_w$ | $v_w$ | $v_w$ | $v_w$ |
| | | | | | (plf) | (kips/in.) | (plf) | (kips/in.) | (plf) | (kips/in.) | (plf) | (kips/in.) | (plf) | (plf) | (plf) | (plf) |
| Structural I | 6d | 1-1/4 | 5/16 | 2 | 370 | 12.0 | 500 | 7.5 | 750 | 10.0 | 840 | 15.0 | 520 | 700 | 1050 | 1175 |
| | | | | 3 | 420 | 9.5 | 560 | 6.0 | 840 | 8.5 | 950 | 13.0 | 590 | 785 | 1175 | 1330 |
| | 8d | 1-3/8 | 3/8 | 2 | 540 | 11.0 | 720 | 7.5 | 1060 | 10.0 | 1200 | 15.0 | 755 | 1010 | 1485 | 1680 |
| | | | | 3 | 600 | 10.0 | 800 | 6.5 | 1200 | 9.0 | 1350 | 13.0 | 840 | 1120 | 1680 | 1890 |
| | 10d | 1-1/2 | 15/32 | 2 | 640 | 17.0 | 850 | 12.0 | 1280 | 15.0 | 1460 | 21.0 | 895 | 1190 | 1790 | 2045 |
| | | | | 3 | 720 | 15.0 | 960 | 9.5 | 1440 | 13.0 | 1640 | 18.0 | 1010 | 1345 | 2015 | 2295 |
| Sheathing and Single-Floor | 6d | 1-1/4 | 5/16 | 2 | 340 | 10.0 | 450 | 7.0 | 670 | 9.5 | 760 | 13.0 | 475 | 630 | 940 | 1065 |
| | | | | 3 | 380 | 9.0 | 500 | 6.0 | 760 | 8.0 | 860 | 12.0 | 530 | 700 | 1065 | 1205 |
| | | | 3/8 | 2 | 370 | 9.5 | 500 | 6.0 | 750 | 8.0 | 840 | 12.0 | 520 | 700 | 1050 | 1175 |
| | | | | 3 | 420 | 8.0 | 560 | 5.0 | 840 | 7.0 | 950 | 10.0 | 590 | 785 | 1175 | 1330 |
| | 8d | 1-3/8 | 3/8 | 2 | 480 | 11.0 | 640 | 7.5 | 960 | 9.5 | 1090 | 13.0 | 670 | 895 | 1345 | 1525 |
| | | | | 3 | 540 | 9.5 | 720 | 6.0 | 1080 | 8.5 | 1220 | 12.0 | 755 | 1010 | 1510 | 1710 |
| | | | 7/16 | 2 | 510 | 10.0 | 680 | 7.0 | 1010 | 9.5 | 1150 | 13.0 | 715 | 950 | 1415 | 1610 |
| | | | | 3 | 570 | 9.0 | 760 | 6.0 | 1140 | 8.0 | 1290 | 12.0 | 800 | 1065 | 1595 | 1805 |
| | | | 15/32 | 2 | 540 | 9.5 | 720 | 6.5 | 1060 | 8.5 | 1200 | 13.0 | 755 | 1010 | 1485 | 1680 |
| | | | | 3 | 600 | 8.5 | 800 | 5.5 | 1200 | 7.5 | 1350 | 11.0 | 840 | 1120 | 1680 | 1890 |
| | 10d | 1-1/2 | 15/32 | 2 | 580 | 15.0 | 770 | 11.0 | 1150 | 14.0 | 1310 | 18.0 | 810 | 1080 | 1610 | 1835 |
| | | | | 3 | 650 | 14.0 | 860 | 9.5 | 1300 | 12.0 | 1470 | 16.0 | 910 | 1205 | 1820 | 2060 |
| | | | 19/32 | 2 | 640 | 14.0 | 850 | 9.5 | 1280 | 12.0 | 1460 | 17.0 | 895 | 1190 | 1790 | 2045 |
| | | | | 3 | 720 | 12.0 | 960 | 8.0 | 1440 | 11.0 | 1640 | 15.0 | 1010 | 1345 | 2015 | 2295 |

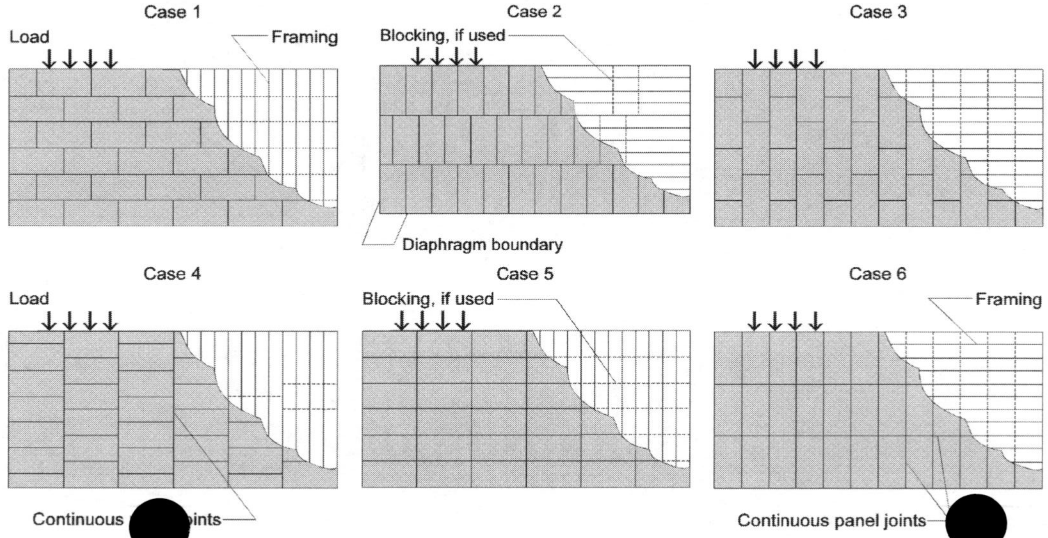

1. Nominal unit shear capacities shall be adjusted in accordance with 4.2.3 to determine ASD allowable unit shear capacity and LRFD factored unit resistance. For general construction requirements see 4.2.6. For specific requirements, see 4.2.7.1 for wood structural panel diaphragms.
2. For framing grades other than Douglas Fir-Larch or Southern Pine, reduced nominal unit shear capacities shall be determined by multiplying the tabulated nominal unit shear capacity by the Specific Gravity Adjustment Factor = [1 – (0.5 – G)], where G = Specific Gravity of the framing lumber from the *NDS*. The Specific Gravity Adjustment Factor shall not be greater than 1.
3. Apparent shear stiffness values, $G_a$, are based on nail slip in framing and panel stiffness values for diaphragms constructed with 3-ply plywood with moisture content less than or equal to 19% at time of fabrication. When 4-ply, 5-ply, or COM-PLY are used, $G_a$ values shall be permitted to be increased by 1.2.
4. Where moisture content of the framing is greater than 19% at time of fabrication, $G_a$ values shall be multiplied by 0.5.

Case 1
Load
Framing

Case 2
Blocking, if used
Diaphragm boundary

Case 3

Case 4
Load
Continuous ints

Case 5
Blocking, if used

Case 6
Framing
Continuous panel joints

# Table A.4.2B Nominal Unit Shear Capacities for Wood-Frame Plywood Diaphragms

## Unblocked Wood Structural Panel Diaphragms[1,2,3,4]

| Sheathing Grade | Common Nail Size | Minimum Fastener Penetration in Framing (in.) | Minimum Nominal Panel Thickness (in.) | Minimum Nominal Framing Width | A SEISMIC — Edge Nail Spacing: 6 in. — Case 1 — $v_s$ (plf) | Case 1 — $G_a$ (kips/in.) | Cases 2,3,4,5,6 — $v_s$ (plf) | Cases 2,3,4,5,6 — $G_a$ (kips/in.) | B WIND — Edge Nail Spacing: 6 in. — Case 1 — $v_w$ (plf) | Cases 2,3,4,5,6 — $v_w$ (plf) |
|---|---|---|---|---|---|---|---|---|---|---|
| Structural I | 6d | 1-1/4 | 5/16 | 2 | 330 | 7.0 | 250 | 4.5 | 460 | 350 |
| | | | | 3 | 370 | 6.0 | 280 | 4.0 | 520 | 390 |
| | 8d | 1-3/8 | 3/8 | 2 | 480 | 7.0 | 360 | 4.5 | 670 | 505 |
| | | | | 3 | 530 | 6.0 | 400 | 4.0 | 740 | 560 |
| | 10d | 1-1/2 | 15/32 | 2 | 570 | 10.0 | 430 | 7.0 | 800 | 600 |
| | | | | 3 | 640 | 9.0 | 480 | 6.0 | 895 | 670 |
| Sheathing and Single-Floor | 6d | 1-1/4 | 5/16 | 2 | 300 | 6.5 | 220 | 4.0 | 420 | 310 |
| | | | | 3 | 340 | 5.5 | 250 | 3.5 | 475 | 350 |
| | | | 3/8 | 2 | 330 | 5.5 | 250 | 4.0 | 460 | 350 |
| | | | | 3 | 370 | 4.5 | 280 | 3.0 | 520 | 390 |
| | 8d | 1-3/8 | 3/8 | 2 | 430 | 6.5 | 320 | 4.5 | 600 | 450 |
| | | | | 3 | 480 | 5.5 | 360 | 3.5 | 670 | 505 |
| | | | 7/16 | 2 | 460 | 6.0 | 340 | 4.0 | 645 | 475 |
| | | | | 3 | 510 | 5.5 | 380 | 3.5 | 715 | 530 |
| | | | 15/32 | 2 | 480 | 5.5 | 360 | 4.0 | 670 | 505 |
| | | | | 3 | 530 | 5.0 | 400 | 3.5 | 740 | 560 |
| | 10d | 1-1/2 | 15/32 | 2 | 510 | 9.0 | 380 | 6.0 | 715 | 530 |
| | | | | 3 | 580 | 8.0 | 430 | 5.5 | 810 | 600 |
| | | | 19/32 | 2 | 570 | 8.5 | 430 | 5.5 | 800 | 600 |
| | | | | 3 | 640 | 7.5 | 480 | 5.0 | 895 | 670 |

1. Nominal unit shear capacities shall be adjusted in accordance with 4.2.3 to determine ASD allowable unit shear capacity and LRFD factored unit resistance. For general construction requirements see 4.2.6. For specific requirements, see 4.2.7.1 for wood structural panel diaphragms.
2. For framing grades other than Douglas Fir-Larch or Southern Pine, reduced nominal unit shear capacities shall be determined by multiplying the tabulated nominal unit shear capacity by the Specific Gravity Adjustment Factor = $[1 - (0.5 - G)]$, where G = Specific Gravity of the framing lumber from the *NDS*. The Specific Gravity Adjustment Factor shall not be greater than 1.
3. Apparent shear stiffness values, $G_a$, are based on nail slip in framing and panel stiffness values for diaphragms constructed with 3-ply plywood with moisture content less than or equal to 19% at time of fabrication. When 4-ply, 5-ply, or COM-PLY are used, $G_a$ values shall be permitted to be increased by 1.2.
4. Where moisture content of the framing is greater than 19% at time of fabrication, $G_a$ values shall be multiplied by 0.5.

Case 1
Load
Framing

Case 2
Blocking, if used
Diaphragm boundary

Case 3

Case 4
Load
Continuous panel joints

Case 5
Blocking, if used

Case 6
Framing
Continuous panel joints

## Table A.4.3A Nominal Unit Shear Capacities for Wood-Frame Plywood Shear Walls[1,2,3]

| Sheathing Material | Minimum Nominal Panel Thickness (in.) | Minimum Fastener Penetration in Framing (in.) | Fastener Type & Size | A SEISMIC Panel Edge Fastener Spacing (in.) | | | | | | | | B WIND Panel Edge Fastener Spacing (in.) | | | |
|---|---|---|---|---|---|---|---|---|---|---|---|---|---|---|---|
| | | | | 6 | | 4 | | 3 | | 2 | | 6 | 4 | 3 | 2 |
| | | | | $v_s$ | $G_a$ | $v_s$ | $G_a$ | $v_s$ | $G_a$ | $v_s$ | $G_a$ | $v_w$ | $v_w$ | $v_w$ | $v_w$ |
| | | | | (plf) | (kips/in.) | (plf) | (kips/in.) | (plf) | (kips/in.) | (plf) | (kips/in.) | (plf) | (plf) | (plf) | (plf) |
| | | | Nail (common or galvanized box) | | | | | | | | | | | | |
| Wood Structural Panels – Structural I | 5/16 | 1-1/4 | 6d | 400 | 10.0 | 600 | 13.0 | 780 | 16.0 | 1020 | 22.0 | 560 | 840 | 1090 | 1430 |
| | 3/8[4] | 1-3/8 | 8d | 460 | 14.0 | 720 | 17.0 | 920 | 20.0 | 1220 | 24.0 | 645 | 1010 | 1290 | 1710 |
| | 7/16[4] | | | 510 | 13.0 | 790 | 16.0 | 1010 | 19.0 | 1340 | 24.0 | 715 | 1105 | 1415 | 1875 |
| | 15/32 | | | 560 | 11.0 | 860 | 14.0 | 1100 | 17.0 | 1460 | 23.0 | 785 | 1205 | 1540 | 2045 |
| | 15/32 | 1-1/2 | 10d | 680 | 16.0 | 1020 | 20.0 | 1330 | 22.0 | 1740 | 28.0 | 950 | 1430 | 1860 | 2435 |
| Wood Structural Panels – Sheathing | 5/16 | 1-1/4 | 6d | 360 | 9.5 | 540 | 12.0 | 700 | 14.0 | 900 | 18.0 | 505 | 755 | 980 | 1260 |
| | 3/8 | | | 400 | 8.5 | 600 | 11.0 | 780 | 13.0 | 1020 | 17.0 | 560 | 840 | 1090 | 1430 |
| | 3/8[4] | 1-3/8 | 8d | 440 | 12.0 | 640 | 15.0 | 820 | 17.0 | 1060 | 20.0 | 615 | 895 | 1150 | 1485 |
| | 7/16[4] | | | 480 | 11.0 | 700 | 14.0 | 900 | 17.0 | 1170 | 21.0 | 670 | 980 | 1260 | 1640 |
| | 15/32 | | | 520 | 10.0 | 760 | 13.0 | 980 | 15.0 | 1280 | 20.0 | 730 | 1065 | 1370 | 1790 |
| | 15/32 | 1-1/2 | 10d | 620 | 14.0 | 920 | 17.0 | 1200 | 19.0 | 1540 | 23.0 | 870 | 1290 | 1680 | 2155 |
| | 19/32 | | | 680 | 13.0 | 1020 | 16.0 | 1330 | 18.0 | 1740 | 22.0 | 950 | 1430 | 1860 | 2435 |

1. Nominal unit shear capacities shall be adjusted in accordance with 4.3.3 to determine ASD allowable unit shear capacity and LRFD factored unit resistance. For general construction requirements see 4.3.6. For specific requirements, see 4.3.7.1 for wood structural panel shear walls.
2. For framing grades other than Douglas Fir-Larch or Southern Pine, reduced nominal unit shear capacities shall be determined by multiplying the tabulated nominal unit shear capacity by the Specific Gravity Adjustment Factor = [1 – (0.5 – G)], where G = Specific Gravity of the framing lumber from the *NDS*. The Specific Gravity Adjustment Factor shall not be greater than 1.
3. Apparent shear stiffness values, $G_a$, are based on nail slip in framing and panel stiffness values for shear walls constructed with 3-ply plywood with moisture content less than or equal to 19% at time of fabrication. When 4-ply, 5-ply, or COM-PLY are used, $G_a$ values shall be permitted to be increased by 1.2. Where moisture content of the framing is greater than 19% at time of fabrication, $G_a$ values shall be multiplied by 0.5.
4. Shears are permitted to be increased to values shown for 15/32 inch sheathing with same nailing provided (a) studs are spaced a maximum of 16 inches on center, or (b) if panels are applied with long dimension across studs.

# REFERENCES

R

AMERICAN FOREST & PAPER ASSOCIATION

# References


1. ASD/LRFD Manual for Engineered Wood Construction, American Forest & Paper Association, Washington, DC, 2005.
2. AHA A194.1-85, Cellulosic Fiber Board, American Hardboard Association, Palatine, IL, 1985.
3. ANSI/AHA A135.4-95, Basic Hardboard, American Hardboard Association, Palatine, IL, 1995.
4. ANSI/AHA A135.5-95, Prefinished Hardboard Paneling, American Hardboard Association, Palatine, IL, 1995.
5. ANSI A208.1-93, Particleboard, ANSI, New York, NY, 1993.
6. ASTM C 28/C 28M-01, Standard Specification for Gypsum Plasters, ASTM, West Conshocken, PA, 2001.
7. ASTM C 36/C 36M-01, Standard Specification for Gypsum Wallboard, ASTM, West Conshocken, PA, 2001.
8. ASTM C 37/C 37M-01, Standard Specification for Gypsum Lath, ASTM, West Conshocken, PA, 2001.
9. ASTM C 79/C 79M-01, Standard Specification for Treated Core and Non-treated Core Gypsum Sheathing Board, ASTM, West Conshocken, PA, 2001.
10. ASTM C 150-00, Standard Specification for Portland Cement, ASTM, West Conshocken, PA, 2000.
11. ASTM C 208-95(2001), Standard Specification for Cellulosic Fiber Insulation Board, ASTM, West Conshocken, PA, 2001.
12. ASTM C 588/C 588M-01, Standard Specification for Gypsum Base for Veneer Plasters, ASTM, West Conshocken, PA, 2001.
13. ASTM C 630/C 630M-01, Standard Specification for Water-Resistant Gypsum Backing Board, ASTM, West Conshocken, PA, 2001.
14. ASTM C 840-01, Standard Specification for Application and Finishing of Gypsum Board, ASTM, West Conshocken, PA, 2001.
15. ASTM C 841-99, Standard Specification for Installation of Interior Lathing and Furring, ASTM, West Conshocken, PA, 1999.
16. ASTM C 844-99, Standard Specification for Application of Gypsum Base to Receive Gypsum Veneer Plaster, ASTM, West Conshocken, PA, 1999.
17. ASTM C 847-95, Standard Specification for Metal Lath, ASTM, West Conshocken, PA, 2000.
18. ASTM C 926-98a, Standard Specification for Application of Portland Cement Based Plaster, ASTM, West Conshocken, PA, 1998.
19. ASTM C 1032-96, Standard Specification for Woven Wire Plaster Base, ASTM, West Conshocken, PA, 1996.
20. ASTM C 1063-99, Standard Specification for Installation of Lathing and Furring to Receive Interior and Exterior Portland Cement-Based Plaster, ASTM, West Conshocken, PA, 1999.
21. ASTM C 1280-99, Standard Specification for Application of Gypsum Sheathing, ASTM, West Conshocken, PA, 1999.
22. National Design Specification (NDS) for Wood Construction, American Forest & Paper Association, Washington, DC, 2005.
23. PS 1-95 Construction and Industrial Plywood, United States Department of Commerce, National Institute of Standards and Technology, Gaithersburg, MD, 1995.
24. PS 2-92 Performance Standard for Wood-Based Structural Use Panels, United States Department of Commerce, National Institute of Standards and Technology, Gaithersburg, MD, 1992.

# SDPWS COMMENTARY



C

AMERICAN FOREST & PAPER ASSOCIATION

# FOREWORD

The *Special Design Provisions for Wind and Seismic (SDPWS)* document was first issued in 2002. It contains provisions for materials, design, and construction of wood members, fasteners, and assemblies to resist wind and seismic forces. The 2005 edition is the second edition of this publication.

The Commentary to the *SDPWS* is provided herein and includes background information for each section as well as sample calculations for each of the design value tables.

The Commentary follows the same subject matter organization as the *SDPWS*. Discussion of a particular provision in the *SDPWS* is identified in the Commentary by the same section or subsection. When available, references to more detailed information on specific subjects are included.

In developing the provisions of the *SDPWS*, data and experience with structures in-service has been carefully evaluated by the AF&PA Wood Design Standards Committee for the purpose of providing a standard of practice. It is intended that this document be used in conjunction with competent engineering design, accurate fabrication, and adequate supervision of construction. Therefore AF&PA does not assume any responsibility for error or omission in the *SDPWS* and *SDPWS Commentary*, nor for engineering designs and plans prepared from it.

Inquiries, comments, and suggestions from the readers of this document are invited.

American Forest & Paper Association

# C2 GENERAL DESIGN REQUIREMENTS

## C2.1 General

### C2.1.1 Scope

Allowable stress design (ASD) and load and resistance factor design (LRFD) provisions are applicable for the design of wood members and systems to resist wind and seismic loads. For other than short-term wind and seismic loads (10-minute basis), adjustment of design capacities for load duration or time effect shall be in accordance with the *National Design Specification® (NDS®) for Wood Construction* (6).

### C2.1.2 Design Methods

Both ASD and LRFD (also referred to as strength design) formats are addressed by reference to the *National Design Specification (NDS) for Wood Construction* (6) for design of wood members and connections. The design of elements throughout a structure will generally utilize either the ASD or LRFD format; however, specific requirements to use a single design format for all elements within a structure are not included. The suitability of mixing formats within a structure is the responsibility of the designer in compliance with requirements of the authority having jurisdiction. *ASCE 7 – Minimum Design Loads for Buildings and Other Structures* (5) limits mixing of design formats to cases where there are changes in materials.

## C2.2 Terminology

**ASD Reduction Factor:** This term denotes the specific adjustment factor used to convert nominal design values to ASD design values.

**Nominal Strength:** Nominal strength (or nominal capacity) is used to provide a common reference point from which to derive ASD design values or LRFD design values. For wood structural panels, tabulated nominal unit shear capacities for wind, $v_w$, (nominal strength) were derived using ASD design values from industry design documents and model building codes (2, 18, 19, 20) times a factor of 2.8. The factor of 2.8, based on minimum performance requirements (8), has commonly been considered the minimum safety factor associated with ASD unit shear capacity for wood structural panel shear walls and diaphragms. For consistency with the ratio of wind and seismic design capacities for wood structural panel shear walls and diaphragms in the model building codes (2), the nominal unit shear capacity for seismic, $v_s$, was derived by dividing the nominal unit shear capacity for wind by 1.4. For fiberboard and lumber shear walls and lumber diaphragms, similar assumptions were used.

For shear walls utilizing other materials, the ASD unit shear capacity values from model building codes (2) and industry design documents (20) were multiplied by 2.0 to develop the nominal unit shear capacity values for both wind and seismic.

**Resistance Factor:** For LRFD, resistance factors are assigned to various wood properties with only one factor for each stress mode (i.e., bending, shear, compression, tension, and stability). Theoretically, the magnitude of a resistance factor is considered to, in part, reflect relative variability of wood product properties. However, for wood design provisions, actual differences in product variability are already embedded in the reference design values. This is due to the fact that typical reference design values are based on a statistical estimate of a near-minimum value (5th percentile).

The following resistance factors are used in the *SDPWS*: a) sheathing in-plane shear, $\phi_D = 0.80$, b) sheathing out-of-plane bending $\phi_b = 0.85$. LRFD resistance factors have been determined by an ASTM consensus standard committee (16). The factors were derived to achieve a target reliability index, $\beta$, of 2.4 for a reference design condition. Examination of other design conditions verified a reasonable range of reliability indices would be achieved by application of ASTM D 5457 (16) resistance factors. Because the target reliability index was selected based on historically acceptable design practice, there is virtually no difference between ASD and LRFD designs at the reference design condition. However, differences will occur due to varying ASD and LRFD load factors and under certain load combinations. It should be noted that this practice (of calibrating LRFD to historically acceptable design) was also used by the other major building materials. The calibration calculation between ASD and LRFD for in-plane shear considered the following:

### Wind Design

$$\text{ASD: } \frac{R_{wind}}{2.0} \geq 1.0W \qquad \text{(C2.2-1)}$$

$$\text{LRFD: } \phi_D R_{wind} \geq 1.6\ W \qquad \text{(C2.2-2)}$$

### Seismic Design

$$\text{ASD: } \frac{R_{seismic}}{2.0} \geq 0.7E \qquad \text{(C2.2-3)}$$

$$\text{LRFD: } \phi_D R_{seismic} \geq 1.0E \qquad \text{(C2.2-4)}$$

$R_{wind}$ = nominal capacity for wind

$R_{seismic}$ = nominal capacity for seismic

2.0 = ASD Reduction Factor

$\phi_D$ = resistance factor for in-plane shear of shear walls and diaphragms

W = wind load effect

E = earthquake load effect

From Equation C2.2-1 and Equation C2.2-2, the value of $\phi_D$ that produces exact calibration between ASD and LRFD design for wind is:

$$\phi_D = \frac{1.6W}{R_{wind}} = \frac{1.6W}{1.0W(2.0)} = 0.80 \qquad \text{(C2.2-5)}$$

From Equation C2.2-3 and Equation C2.2-4, the value of $\phi_D$ that produces exact calibration between ASD and LRFD design for seismic is:

$$\phi_D = \frac{1.0E}{R_{seismic}} = \frac{1.0E}{0.7E(2.0)} = 0.70 \qquad \text{(C2.2-6)}$$

A single resistance factor, $\phi_D$, of 0.80 for wind and seismic design was chosen by both the ASTM and the *SDPWS* consensus committees because the added complexity of utilizing two separate factors was not warranted given the small relative difference in calibrations. The same approach was used for earlier calibrations and resulted in $\phi_D = 0.65$ as shown in *ASCE* 16-95 and the 2001 *SDPWS*; however, the calibration was tied to load combinations given in *ASCE* 7-88 resulting in a value of $\phi_D = 0.65$.

Recalling that nominal unit shear capacities for seismic were derived by dividing the nominal unit shear capacity for wind by 1.4 (see C2.2 Nominal Strength), the "Effective $\phi_D$" for seismic shear resistance is approximately 0.57:

$$\text{"Effective } \phi_D\text{"} = \frac{0.80}{1.4} = 0.57 \qquad \text{(C2.2-7)}$$

**where:**

0.80 = $\phi_D$ from Equation C2.2-5 calibration for wind

1.4 = ratio of $R_{wind}$ to $R_{seismic}$ ($R_{wind}/R_{seismic}$)

From Equation C2.2-7, the LRFD factored unit shear resistance for seismic is approximately 0.57 times the minimum target strength (e.g., $R_{wind}$) set by underlying product standards.

# C3 MEMBERS AND CONNECTIONS

## C3.1 Framing

### C3.1.1 Wall Framing

Wall studs sheathed on both sides are stronger and stiffer in flexure (i.e., wind loads applied perpendicular to the wall plane) than those in similar, unsheathed wall assemblies. The enhanced performance or "system effect" is recognized in wood design with the repetitive member factor, $C_r$, which accounts for effects of partial composite action and load-sharing (1). Wall stud bending stress increase factors in *SDPWS* Table 3.1.1.1 are applicable for out-of-plane wind loads and were derived based on wall tests (9). A factor of 1.56 was determined for a wall configured as follows:

| | |
|---|---|
| **Framing** | 2x4 Stud grade Douglas fir studs at 16" o.c. |
| **Interior Sheathing** | ½" gypsum wallboard attached with 4d cooler nails at 7" o.c. edge and 10" o.c. field (applied vertically). |
| **Exterior Sheathing** | 3/8" rough sanded 303 siding attached with 6d box nails at 6" o.c. edge and 12" o.c. field (applied vertically). |

For design purposes, a slightly more conservative value of 1.5 was chosen to represent a modified 2x4 stud wall system as follows:

| | |
|---|---|
| **Framing** | 2x4 Stud grade Douglas fir studs at 24" o.c. |
| **Interior Sheathing** | ½" gypsum wallboard attached with 5d cooler nails at 7" o.c. edge and 10" o.c. field (applied vertically). |
| **Exterior Sheathing** | 3/8" wood structural panels attached with 8d common nails at 6" o.c. edge and 12" o.c. field (blocked). |

For other stud depths, the wall stud bending stress increase factor was assumed to be proportional to the relative stiffness (EI) of the stud material. A repetitive member factor of 1.15 (6) was assumed for a 2x12 stud in a wall system and Equation C3.1.1-1 was used to interpolate repetitive member factors for 2x6, 2x8, and 2x10 studs:

$$C_r = 1.15\left[\frac{178in^4}{I_{stud}}\right]^{0.076} \quad \text{(C3.1.1-1)}$$

Slight differences between calculated $C_r$ values and those appearing in *SDPWS* Table 3.1.1.1 are due to rounding.

## C3.2 Sheathing

Nominal uniform load capacities in *SDPWS* Tables 3.2.1 and 3.2.2 assume a two-span continuous condition. Out-of-plane sheathing capacities are often tabulated in other documents on the basis of a three-span continuous condition. Although the three-span continuous condition results in higher capacity, the more conservative two-span continuous condition was selected because this condition frequently exists at building end zones where the largest wind forces occur.

Examples C3.2.1-1 and C3.2.1-2 illustrate how the values in *SDPWS* Table 3.2.1 were generated using wood structural panel out-of-plane bending and shear values given in Tables C3.2A and C3.2B. Although the following two examples are for *SDPWS* Table 3.2.1, the same procedure can be used to generate the values shown in *SDPWS* Table 3.2.2.

Table C3.2C provides out-of-plane bending strength capacities for cellulosic fiberboard sheathing based on minimum modulus of rupture criteria in ASTM C208. Values in *SDPWS* Table 3.2.2 can be derived using the same procedure as described in Example C3.2.1-1.

### Table C3.2A Wood Structural Panel Dry Design Bending Strength Capacities

| Span Rating: *Sheathing* | Bending Strength, $F_b S$ (lb-in./ft width) | |
|---|---|---|
| | Strength Axis Perpendicular to Supports | Strength Axis Parallel to Supports |
| 24/0 | 250 | 54 |
| 24/16 | 320 | 64 |
| 32/16 | 370 | 92 |
| 40/20 | 625 | 150 |
| 48/24 | 845 | 225 |

### Table C3.2B Wood Structural Panel Dry Shear Capacities in the Plane

| Span Rating: *Sheathing* | Shear in the Plane, $F_S$ [lb/Q] (lb/ft width) |
|---|---|
| | Strength Axis Either Perpendicular or Parallel to Supports |
| 24/0 | 130 |
| 24/16 | 150 |
| 32/16 | 165 |
| 40/20 | 205 |
| 48/24 | 250 |

### Table C3.2C Cellulosic Fiberboard Sheathing Design Bending Strength Capacities

| Span Rating: *Sheathing* | Bending Strength, $F_b S$ (lb-in./ft width) |
|---|---|
| | Strength Axis Either Parallel or Perpendicular to Supports |
| Regular 1/2" | 55 |
| Structural 1/2" | 80 |
| Structural 25/32" | 97 |

## EXAMPLE C3.2.1-1 Determine the Nominal Uniform Load Capacity in SDPWS Table 3.2.1

Determine the nominal uniform load capacity in *SDPWS* Table 3.2.1 Nominal Uniform Load Capacities (psf) for Wall Sheathing Resisting Out-of-Plane Wind Loads for the following conditions:

| | |
|---|---|
| Sheathing type | = wood structural panels |
| Span rating or grade | = 24/0 |
| Min. thickness | = 3/8 in. |
| Strength axis | = perpendicular to supports |
| Actual stud spacing | = 12 in. |

ASD (normal load duration, i.e., 10-yr) bending capacity:

$F_b S$ = 250 lb-in./ft width from Table C3.2A

ASD (normal load duration, i.e., 10-yr) shear capacity:

$F_s I b/Q$ = 130 lb/ft width from Table C3.2B

Maximum uniform load based on bending strength for a two-span condition:

$$w_b = \frac{96F_bS}{l^2} = \frac{96 \times 250}{12^2} = 167 \text{ psf}$$

Maximum uniform load based on shear strength for a two-span condition:

$$w_s = \frac{19.2F_sIb/Q}{l_{clearspan}} = \frac{19.2 \times 130}{(12-1.5)} = 238 \text{ psf}$$

Maximum uniform load based on bending governs. Converting to the nominal capacity basis of *SDPWS* Table 3.2.1:

$$w_{\text{nominal}} = \left(\frac{2.16}{\phi_b}\right) \times ASD_{10\text{-}yr}$$

*SDPWS* Table 3.2.1

$$= \frac{2.16}{0.85} \times 167 = 424 \text{ psf}$$

$$\approx 425 \text{ psf}$$

**where:**

2.16/0.85 = conversion from a normal load duration (i.e., 10-yr ASD basis) to the short-term (10-min) nominal capacity basis of *SDPWS* Table 3.2.1.

## EXAMPLE C3.2.1-2 Determine the Nominal Uniform Load Capacity in SDPWS Table 3.2.1

Determine the nominal uniform load capacity in *SDPWS* Table 3.2.1 Nominal Uniform Load Capacities (psf) for Wall Sheathing Resisting Out-of-Plane Wind Loads for the following conditions:

| | |
|---|---|
| Sheathing type | = wood structural panels |
| Span rating or grade | = 40/20 |
| Min. thickness | = 19/32 in. |
| Strength axis | = perpendicular to supports |
| Actual stud spacing | = 12 in. |

ASD (normal load duration, i.e., 10-yr) bending capacity:

$F_b\,S$ = 625 lb-in./ft width from Table C3.2A

ASD (normal load duration, i.e., 10-yr) shear capacity:

$F_s\,I\,b/Q$ = 205 lb/ft width from Table C3.2B

Maximum uniform load based on bending strength for a two-span condition:

$$w_b = \frac{96F_bS}{l^2} = \frac{96\times 625}{12^2} = 417 \text{ psf}$$

Maximum uniform load based on shear strength for a two-span condition:

$$w_s = \frac{19.2F_sIb/Q}{l_{clearspan}} = \frac{19.2\times 205}{(12-1.5)} = 375 \text{ psf}$$

Maximum uniform load based on shear governs. Converting to the nominal capacity basis of *SDPWS* Table 3.2.1:

$$w_{nominal} = \left(\frac{2.16}{\phi_b}\right)\times ASD_{10\text{-}yr}$$

*SDPWS* Table 3.2.1

$$= \frac{2.16}{0.85}\times 375 = 953 \text{ psf}$$

$$\approx 955 \text{ psf}$$

**where:**

2.16/0.85 = conversion from a normal load duration (i.e., 10-yr ASD basis) to the short-term (10-min) nominal capacity basis of *SDPWS* Table 3.2.1.

# C3.3 Connections

Section 3.3 refers the user to the *NDS* (6) when designing connections to resist wind or seismic forces. In many cases, resistance to out-of-plane forces due to wind may be limited by connection capacity (withdrawal capacity of the connection) rather than out-of-plane bending or shear capacity of the panel.

# C4 LATERAL FORCE-RESISTING SYSTEMS

## C4.1 General

### C4.1.1 Design Requirements

General design requirements for lateral force-resisting systems are described in this section and are applicable to engineered structures.

### C4.1.2 Shear Capacity

Nominal unit shear capacities (see C2.2) for wind and seismic require adjustment in accordance with *SDPWS* 4.2.3 for diaphragms and *SDPWS* 4.3.3 for shear walls to derive an appropriate design value.

### C4.1.3 Deformation Requirements

Consideration of deformations (such as deformation of the overall structure, elements, connections, and systems within the structure) that can occur is necessary to maintain load path and ensure proper detailing. Special requirements are provided for wood members resisting forces from concrete and masonry (see C4.1.5) due to potentially large differences in stiffness and deflection limits for wood and concrete systems as well as open front buildings (see C4.2.5.1.1) where forces are distributed by diaphragm rotation.

### C4.1.4 Boundary Elements

Boundary elements must be sized to transfer the design tension and compression forces. Good construction practice and efficient design and detailing for boundary elements utilizes framing members in the plane or tangent to the plane of the diaphragm or shear wall.

### C4.1.5 Wood Members and Systems Resisting Seismic Forces Contributed by Masonry and Concrete Walls

The use of wood diaphragms with masonry or concrete walls is common practice. Story height and other limitations for wood members and wood systems resisting seismic forces from concrete or masonry walls are given to address deformation compatibility and are largely based on field observations following major seismic events. Due to significant differences in stiffness, wood diaphragms and horizontal trusses are not permitted where forces contributed by masonry or concrete walls results in torsional force distribution through the diaphragm or truss.

The term "horizontal trusses" refers to trusses that are oriented such that their top and bottom chords and web members are in the plane of the lateral load and resist those lateral loads. In this context, a horizontal truss is a bracing system capable of resisting horizontal seismic forces contributed by masonry or concrete walls.

Where wood structural panel shear walls are used to provide resistance to seismic forces contributed by masonry and concrete walls, deflections are limited to 0.7% of the story height in accordance with deflection limits (5) for masonry and concrete construction. The intent is to limit failure of the masonry or concrete portions of the structure due to excessive deflection.

### C4.1.6 Wood Members and Systems Resisting Seismic Forces from Other Concrete or Masonry Construction

Seismic forces from other concrete or masonry construction (i.e., other than walls) are permitted and should be accounted for in design. *SDPWS* 4.1.6 is not intended to restrict the use of concrete floors – including wood floors with concrete toppings as well as reinforced concrete slabs – or similar such elements in floor construction. It is intended to clarify that, where such elements are present in combination with a wood system, the wood system shall be designed to account for the seismic forces generated by the additional mass of such elements.

Design of wood members to support the additional mass of concrete and masonry elements shall be in accordance with the *NDS* and required deflection limits as specified in concrete or masonry standards or the model building codes (2).

### C4.1.7 Toe-Nailed Connections

Limits on use of toe-nailed connections in seismic design categories D, E, and F for transfer of seismic forces is consistent with building code requirements (2). Test data (12) suggests that the toe-nailed connection limit on a bandjoist to wall plate connection may be too restrictive; however, an appropriate alternative limit requires further study. Where blocking is used to transfer high seismic forces, toe-nailed connections can sometimes split the block or provide a weakened plane for splitting.

## C4.2 Wood Diaphragms

### C4.2.1 Application Requirements

General requirements for wood diaphragms include consideration of diaphragm strength and deflection.

### C4.2.2 Deflection

The total mid-span deflection of a blocked, uniformly nailed wood structural panel diaphragm can be calculated by summing the effects of four sources of deflection: framing bending deflection, panel shear deflection, deflection from nail slip, and deflection due to chord splice slip:

*(bending) (shear) (nail slip) (chord slip)*

$$\delta_{dia} = \frac{5vL^3}{8EAW} + \frac{vL}{4G_vt_v} + 0.188Le_n + \frac{\sum(x\Delta_c)}{2W} \quad \text{(C4.2.2-1)}$$

**where:**

v = induced unit shear, plf

L = diaphragm dimension perpendicular to the direction of the applied force, ft

E = modulus of elasticity of diaphragm chords, psi

A = area of chord cross-section, in.$^2$

W = width of diaphragm in direction of applied force, ft

$G_vt_v$ = shear stiffness, lb/in. of panel depth. See Table C4.2.2A or C4.2.2B.

x = distance from chord splice to nearest support, ft

$\Delta_c$ = diaphragm chord splice slip at the induced unit shear, in.

$e_n$ = nail slip, in. See Table C4.2.2D.

**Table C4.2.2A Shear Stiffness, $G_vt_v$ (lb/in. of depth), for Wood Structural Panels**

| Span Rating | Minimum Nominal Panel Thickness (in.) | Structural Sheathing | | | | Structural I | | | |
|---|---|---|---|---|---|---|---|---|---|
| | | Plywood | | | OSB | Plywood | | | OSB |
| | | 3-ply | 4-ply | 5-ply | | 3-ply | 4-ply | 5-ply | |
| **Sheathing Grades** | | | | | | | | | |
| 24/0 | 3/8 | 25,000 | 32,500 | 37,500 | 77,500 | 32,500 | 42,500 | 41,500 | 77,500 |
| 24/16 | 7/16 | 27,000 | 35,000 | 40,500 | 83,500 | 35,000 | 45,500 | 44,500 | 83,500 |
| 32/16 | 15/32 | 27,000 | 35,000 | 40,500 | 83,500 | 35,000 | 45,500 | 44,500 | 83,500 |
| 40/20 | 19/32 | 28,500 | 37,000 | 43,000 | 88,500 | 37,000 | 48,000 | 47,500 | 88,500 |
| 48/24 | 23/32 | 31,000 | 40,500 | 46,500 | 96,000 | 40,500 | 52,500 | 51,000 | 96,000 |
| **Single Floor Grades** | | | | | | | | | |
| 16 oc | 19/32 | 27,000 | 35,000 | 40,500 | 83,500 | 35,000 | 45,500 | 44,500 | 83,500 |
| 20 oc | 19/32 | 28,000 | 36,500 | 42,000 | 87,000 | 36,500 | 47,500 | 46,000 | 87,000 |
| 24 oc | 23/32 | 30,000 | 39,000 | 45,000 | 93,000 | 39,000 | 50,500 | 49,500 | 93,000 |
| 32 oc | 7/8 | 36,000 | 47,000 | 54,000 | 110,000 | 47,000 | 61,000 | 59,500 | 110,000 |
| 48 oc | 1-1/8 | 50,500 | 65,500 | 76,000 | 155,000 | 65,500 | 85,000 | 83,500 | 155,000 |

1. Sheathing grades used for calculating $G_a$ values for diaphragm and shear wall tables.
2. $G_vt_v$ values for 3/8" panels with span rating of 24/0 used to estimate $G_a$ values for 5/16" panels.
3. 5-ply applies to plywood with five or more layers. For 5-ply plywood with three layers, use $G_vt_v$ values for 4-ply panels.

## Table C4.2.2B Shear Stiffness, $G_v t_v$ (lb/in. of depth), for Other Sheathing Materials

| Sheathing Material | Minimum Nominal Panel Thickness (in.) | $G_v t_v$ |
|---|---|---|
| Plywood Siding | 5/16 & 3/8 | 25,000 |
| Particleboard | 3/8 | 25,000 |
| | 1/2 | 28,000 |
| | 5/8 | 28,500 |
| Fiberboard | 1/2 & 25/32 | 25,000 |
| Gypsum board | 1/2 & 5/8 | 40,000 |
| Lumber | All | 25,000 |

## Table C4.2.2C Relationship Between Span Rating and Nominal Thickness

| Span Rating | Nominal Thickness (in.) | | | | | | | | | | |
|---|---|---|---|---|---|---|---|---|---|---|---|
| | 3/8 | 7/16 | 15/32 | 1/2 | 19/32 | 5/8 | 23/32 | 3/4 | 7/8 | 1 | 1-1/8 |
| **Sheathing** | | | | | | | | | | | |
| 24/0 | P | A | A | A | | | | | | | |
| 24/16 | | P | A | A | | | | | | | |
| 32/16 | | | P | A | A | A | | | | | |
| 40/20 | | | | | P | A | A | A | | | |
| 48/24 | | | | | | | P | A | A | | |
| **Single Floor Grade** | | | | | | | | | | | |
| 16 oc | | | | | P | A | | | | | |
| 20 oc | | | | | P | A | | | | | |
| 24 oc | | | | | | | P | A | | | |
| 32 oc | | | | | | | | | P | A | |
| 48 oc | | | | | | | | | | | P |

P = Predominant nominal thickness for each span rating.
A = Alternative nominal thickness that may be available for each span rating. Check with suppliers regarding availability.

## Table C4.2.2D Fastener Slip, $e_n$ (in.)

| Sheathing | Fastener Size | Maximum Fastener Load ($V_n$) (lb/fastener) | Fastener Slip, $e_n$ (in.) | |
|---|---|---|---|---|
| | | | Fabricated w/green (>19% m.c.) lumber | Fabricated w/dry (≤ 19% m.c.) lumber |
| Wood Structural Panel (WSP) or Particleboard[1] | 6d common | 180 | $(V_n/434)^{2.314}$ | $(V_n/456)^{3.144}$ |
| | 8d common | 220 | $(V_n/857)^{1.869}$ | $(V_n/616)^{3.018}$ |
| | 10d common | 260 | $(V_n/977)^{1.894}$ | $(V_n/769)^{3.276}$ |
| Fiberboard | All | - | - | 0.07 |
| Gypsum Board | All | - | - | 0.03 |
| Lumber | All | - | - | 0.07 |

1. Slip values are based on plywood and OSB fastened to lumber with a specific gravity of 0.50 or greater. The slip shall be increased by 20 percent when plywood is not Structural I. Nail slip for common nails have been extended to galvanized box or galvanized casing nails of equivalent penny weight for purposes of calculating $G_a$.

*SDPWS* Equation 4.2-1 is a simplification of Equation C4.2.2-1, using only three terms for calculation of the total mid-span diaphragm deflection:

$$\delta_{dia} = \underset{(bending)}{\frac{5vL^3}{8EAW}} + \underset{(shear)}{\frac{0.25vL}{1000G_a}} + \underset{(chord\ slip)}{\frac{\sum(x\Delta_c)}{2W}} \quad \text{(C4.2.2-2)}$$

**where:**

v = induced unit shear, plf

L = diaphragm dimension perpendicular to the direction of the applied force, ft

E = modulus of elasticity of diaphragm chords, psi

A = area of chord cross-section, in.²

W = width of diaphragm in direction of applied force, ft

$G_a$ = apparent diaphragm shear stiffness, kips/in.

x = distance from chord splice to nearest support, ft

$\Delta_c$ = diaphragm chord splice slip at the induced unit shear, in.

In Equation C4.2.2-2, panel shear and nail slip are assumed to be inter-related and have been combined into a single term to account for shear deformations. Equation C4.2.2-3 relates apparent shear stiffness, $G_a$, to nail slip and panel shear stiffness:

$$G_a = \frac{1.4v_{s(ASD)}}{\frac{1.4v_{s(ASD)}}{G_v t_v} + 0.75e_n} \quad \text{(C4.2.2-3)}$$

**where:**

$1.4\ v_{s(ASD)}$ = 1.4 times the ASD unit shear capacity for seismic. The value of 1.4 converts ASD level forces to strength level forces.

Calculated deflection, using either the 4-term (Equation C4.2.2-1) or 3-term equation (*SDPWS* Equation 4.2-1), is identical at the critical strength design level — 1.4 times the allowable shear value for seismic (see Figure C4.3.2).

For unblocked wood structural panel diaphragms, tabulated values of $G_a$ are based on limited test data for blocked and unblocked diaphragms (3, 4, 11). For diaphragms of Case 1, reduced shear stiffness equal to $0.6G_a$ was used to derive tabulated $G_a$ values. For unblocked diaphragms of Case 2, 3, 4, 5, and 6, reduced shear stiffness equal to $0.4G_a$ was used to derive tabulated $G_a$ values. Tests of blocked and unblocked diaphragms (4) are compared in Table C4.2.2E for diaphragms constructed as follows:

Sheathing material = Sheathing Grade, 3/8" minimum nominal panel thickness
Nail size = 8d common nail
Diaphragm length = 24 ft
Diaphragm width = 24 ft
Panel edge nail spacing = 6 in.
Boundary nail spacing = 6 in. o.c. at boundary parallel to load (4 in. o.c. at boundary perpendicular to load for walls A and B)

Calculated deflections at 1.4 x $v_{s(ASD)}$ closely match test data for blocked and unblocked diaphragms.

In diaphragm table footnotes, a factor of 0.5 is provided to adjust tabulated $G_a$ values (based on fabricated dry condition) to approximate $G_a$ where "green" framing is used. This factor is based on analysis of apparent shear stiffness for wood structural panel shear wall and diaphragm construction where:

1) framing moisture content is greater than 19% at time of fabrication (green), and
2) framing moisture content is less than or equal to 19% at time of fabrication (dry).

The average ratio of "green" to "dry" for $G_a$ across shear wall and diaphragm cells ranged from approximately 0.52 to 0.55. A rounded value of 0.5 results in slightly greater values of calculated deflection for "green" framing when compared to the more detailed 4-term deflection equations. Although based on nail slip relationships applicable to wood structural panel shear walls, this reduction can also be extended to lumber sheathed diaphragm construction.

In Table C4.2.2F, calculated deflections using *SDPWS* Equation 4.2-1 are compared to deflections from tests at 1.4 times the allowable seismic design value for a horizontally sheathed and single diagonally sheathed lumber diaphragm. Calculated deflections show reasonable agreement (within 3/16") with those from tests (26) of 20 ft x 60 ft (W = 20 ft, L = 60 ft) diaphragms. Calculated deflections include estimates of deflection due to bending, shear, and chord slip. For both diaphragms, calculated shear deformation accounted for nearly 85% of the total calculated mid-span deflection. Tested deflection for Diaphragm 4 is slightly greater than estimated by calculation and may be attributed to limited effectiveness of the diaphragm chord construction which utilized blocking to transfer forces to the double 2x6 top plate chord. For Diaphragm 2, chord construction utilized 2-2x10 bandjoists.

## Table C4.2.2E Data Summary for Blocked and Unblocked Wood Structural Panel Diaphragms

| Wall | Blocked/ Unblocked | $1.4v_{s(ASD)}$ (plf) | Actual Deflection, (in.) | Apparent Stiffness[1], $G_a$, (kips/in.) | Calculated Deflection, (in.) | Diaphragm Layout |
|---|---|---|---|---|---|---|
| A | Blocked | 378 | 0.22 | 14.4 | 0.18 | Case 1 |
| D | Unblocked | 336 | 0.26 | (0.60 x 14.4) = 8.6 | 0.26 | Case 1 |
| B | Blocked | 378 | 0.15 | 14.4 | 0.18 | Case 3 |
| E | Unblocked | 252 | 0.23 | (0.40 x 14.4) = 5.8 | 0.29 | Case 3 |

1. Values of $G_a$ for the blocked diaphragm case were taken from *SDPWS* Table A.4.2A and multiplied by 1.2 (see footnote 3) because sheathing material was assumed to be comparable to 4/5-ply construction.

## Table C4.2.2F Data Summary for Horizontal Lumber and Diagonal Lumber Sheathed Diaphragms

| Diaphragm | Description | Calculated | | | Actual |
|---|---|---|---|---|---|
| | | $1.4v_{s(ASD)}$ (plf) | $G_a$ (kips/in.) | $\delta^1$ (in.) | $\delta$ (in.) |
| Diaphragm 4 | Horizontal Lumber Sheathing<br>– Dry Lumber Sheathing<br>– 2 x 6 chord (double top plates), 5 splices | 70 | 1.5 | 0.81 | 0.93 |
| Diaphragm 2 | Diagonal Lumber Sheathing<br>– Green Lumber Sheathing<br>– 2 x 10 chord, 3 splices<br>– Exposed outdoors for 1 month | 420 | 6.0 | 1.23 | 1.05 |

1. Calculated deflection equal to 0.81" includes estimates of deflection due to bending, shear, and chord slip (0.036" + 0.7" + 0.07" = 0.81"). Calculated deflection equal to 1.23" includes estimates of deflection due to bending, shear, and chord slip (0.13" + 1.05" + 0.05" = 1.23").

## EXAMPLE C4.2.2-1 Derive $G_a$ in SDPWS Table 4.2A

Derive $G_a$ in SDPWS Table 4.2A for a blocked wood structural panel diaphragm constructed as follows:

Sheathing grade = Structural I (OSB)
Common nail size = 6d
Minimum nominal panel thickness = 5/16 in.
Boundary and panel edge nail spacing = 6 in.
Minimum nominal framing width = 2 in.
Nominal unit shear capacity for seismic, $v_s$ = 370 plf *SDPWS* Table 4.2A

Allowable unit shear capacity for seismic:

$v_{s(ASD)}$ = 370 plf/2 = 185 plf

Panel shear stiffness:

$G_v t_v$ = 77,500 lb/in. of panel depth Table C4.2.2A

Nail load/slip at 1.4 $v_s$ (ASD):

$V_n$ = fastener load (lbf/nail)
= 1.4 $v_{s(ASD)}$ (6 in.)/(12 in.)
= 129.5 lb/nail

$e_n = (V_n/456)^{3.144}$ Table C4.2.2D
= (129.5/456)3.144 = 0.0191 in.

**Calculate $G_a$:**

$$G_a = \frac{1.4v_{s(ASD)}}{\frac{1.4v_{s(ASD)}}{G_v t_v} + 0.75e_n} \qquad (C4.2.2\text{-}3)$$

= 14,660 lb/in. ≈ 15 kips/in. *SDPWS* Table 4.2A

C

## EXAMPLE C4.2.2-2 Derive $G_a$ in SDPWS Table 4.2B

Derive $G_a$ in SDPWS Table 4.2B for an unblocked wood structural panel diaphragm constructed as follows:

| | |
|---|---|
| Sheathing grade | = Structural I (OSB) |
| Common nail size | = 6d |
| Minimum nominal panel thickness | = 5/16 in. |
| Minimum nominal framing width | = 2 in. |
| Boundary and panel edge nail spacing | = 6 in. |

$G_a$ = 15 kips/in. *SDPWS* Table 4.2A

**Case 1 - unblocked**

$G_a$ = 0.6 $G_a$ (blocked)

= 0.6 (15.0) = 9.0 kips/in. *SDPWS* Table 4.2B

**Cases 2, 3, 4, 5, and 6 - unblocked**

$G_a$ = 0.4 $G_a$ (blocked)

= 0.4 (15.0) = 6.0 kips/in. *SDPWS* Table 4.2B

## C4.2.3 Unit Shear Capacities

ASD and LRFD unit shear capacities for wind and seismic are calculated as follows from nominal values for wind, $v_w$, and seismic, $v_s$.

ASD unit shear capacity for wind, $v_{w(ASD)}$:

$$v_{w(ASD)} = \frac{v_w}{2.0} \quad \text{(C4.2.3-1)}$$

ASD unit shear capacity for seismic, $v_{s(ASD)}$:

$$v_{s(ASD)} = \frac{v_s}{2.0} \quad \text{(C4.2.3-2)}$$

**where:**

2.0 = ASD reduction factor

LRFD unit shear capacity for wind, $v_{w(LRFD)}$:

$$v_{w(LRFD)} = 0.8 v_w \quad \text{(C4.2.3-3)}$$

LRFD unit shear capacity for seismic, $v_{s(LRFD)}$:

$$v_{s(LRFD)} = 0.8 v_s \quad \text{(C4.2.3-4)}$$

**where:**

0.8 = resistance factor, $\phi_D$, for shear walls and diaphragms

## C4.2.4 Diaphragm Aspect Ratios

Maximum aspect ratios for floor and roof diaphragms (*SDPWS* Table 4.2.4) using wood structural panel or diagonal board sheathing are based on building code requirements (see *SDPWS* 4.2.5.1 for aspect ratio limits for cases where a torsional irregularity exists, for open front structures, and cantilevered diaphragms).

## C4.2.5 Horizontal Distribution of Shear

General seismic design requirements (5) define conditions applicable for the assumption of flexible diaphragms. For flexible diaphragms, the loads should be distributed to wall lines according to tributary area whereas for rigid diaphragms, the loads should be distributed according to relative stiffness.

The distribution of seismic forces to the vertical elements (shear walls) of the seismic-force-resisting system is dependent on: 1) the stiffness of the vertical elements relative to horizontal elements, and 2) the relative stiffness of the various vertical elements.

Where a series of vertical elements of the seismic-force-resisting system are aligned in a row, seismic forces will distribute to the different elements according to their relative stiffness.

C4.2.5.1 Torsional Irregularity: Excessive torsional response of a structure can be a potential cause of failure. As a result, diaphragm dimension and diaphragm aspect ratio limitations are provided for different building configurations. The test for torsional irregularity is consistent with general seismic design requirements (5).

C4.2.5.1.1 Open Front Structures: A structure with shear walls on three sides only (open front) is one category of structure that requires transfer of forces through rotation. Shear force is transferred to shear wall(s) parallel to the applied force and moment due to eccentric loading is transferred into perpendicular walls. Applicable limitations are provided in *SDPWS* 4.2.5.1.1. Both prescriptive limitations on diaphragm length and diaphragm aspect ratio, and requirements of general seismic design criteria (5) including drift limits, increased forces due to presence of irregularities, and increased forces in accordance with redundancy provisions, should be considered in design.

C4.2.5.2 Cantilevered Diaphragms: Limitations on cantilever distance and diaphragm aspect ratios for diaphragms that cantilever horizontally past the outermost shear wall (or other vertical lateral force resisting element) are in addition to requirements of general seismic design criteria (5), including drift limits, increased forces due to presence of irregularities, and increased forces in accordance with redundancy provisions, that should be considered in design.

## C4.2.6 Construction Requirements

C4.2.6.1 Framing Requirements: The transfer of forces into and out of diaphragms is required for a continuous load path. Boundary elements must be sized and connected to the diaphragm to ensure force transfer. This section provides basic framing requirements for boundary elements in diaphragms. Good construction practice and efficient design and detailing for boundary elements utilizes framing members in the plane of the diaphragm or tangent to the plane of the diaphragm (see C4.1.4). Where splices occur in the boundary elements, the transfer of force between the boundary elements should be through the addition of framing members or metal connectors. The use of diaphragm sheathing to splice boundary elements is not permitted.

C4.2.6.2 Sheathing: Sheathing types for diaphragms included in *SDPWS* Table 4.2A and 4.2B are categorized in terms of the following structural use panel grades: Structural I, Sheathing, and Single-Floor. Sheathing grade rated for subfloor, roof, and wall use is usually unsanded and is manufactured with intermediate and exterior glue. The Structural I sheathing grade is used where the greatest available shear and cross-panel strength properties are required. Structural I is made with exterior glue only. The Single-Floor sheathing grade is rated for use as a combination subfloor underlayment, usually with tongue and groove edges, and has sanded or touch sanded faces.

*SDPWS* Table 4.2A and Table 4.2B are applicable to oriented strand board (OSB). *SDPWS* Appendix Tables A.4.2A and A.4.2B are applicable to plywood. While strength properties between equivalent grades and thickness of OSB and plywood are the same, shear stiffness of OSB is greater than that of plywood of equivalent grade and thickness.

C4.2.6.3 Fasteners: Adhesive attachment in diaphragms can only be used in combination with fasteners.

## C4.2.7 Diaphragm Assemblies

C4.2.7.1 Wood Structural Panel Diaphragms: Where wood structural panel sheathing is applied to solid lumber planking or laminated decking – such as in a retrofit or new construction where wood structural panel diaphragm capacities are desired – additional fastening, aspect ratio limits, and other requirements are prescribed to develop diaphragm capacity and transfer forces to boundary elements.

C4.2.7.1.1 Blocked and Unblocked Diaphragms: Standard construction of wood structural panel diaphragms requires use of full size sheets, not less than 4' x 8' except at changes in framing where smaller sections may be needed to cover the roof or floor in question. Unblocked panel widths are limited to 24". Where smaller widths are used, panel edges must be blocked or supported by framing members. The 24" width limit coincides with the minimum width where panel strength capacities for bending and axial tension are applicable (6). For widths less than 24", capacities for bending and axial tension should be reduced in accordance with applicable panel size adjustment factors (panel width adjustment factors are described in the *Commentary to the National Design Specification for Wood Construction* (6)). Apparent shear stiffness values provided in *SDPWS* Table 4.2A and Table 4.2B are based on standard assumptions for panel shear stiffness for oriented strand board (OSB) and nail load slip (see C4.2.2). *SDPWS* Appendix Table A.4.2A and A.4.2B are based on standard assumptions for plywood panel shear stiffness and nail load slip (see C4.2.2).

C4.2.7.2 Diaphragms Diagonally Sheathed with Single-Layer of Lumber: Single diagonally sheathed lumber diaphragms have comparable strength and stiffness to many wood structural panel diaphragm systems. Apparent shear stiffness in *SDPWS* Table 4.2C is based on assumptions of relative stiffness and nail slip (see C4.2.2).

C4.2.7.3 Diaphragms Diagonally Sheathed with Double-Layer of Lumber: Double diagonally sheathed lumber diaphragms have comparable strength and stiffness to many wood structural panel diaphragm systems. Apparent shear stiffness in *SDPWS* Table 4.2C is based on assumptions of relative stiffness and nail slip (see C4.2.2).

C4.2.7.4 Diaphragms Horizontally Sheathed with Single-Layer of Lumber: Horizontally sheathed lumber diaphragms have low strength and stiffness when compared to those provided by wood structural panel diaphragms and diagonally sheathed lumber diaphragms of the same overall dimensions. In new and existing construction, added strength and stiffness can be developed through attachment of wood structural panels over horizontally sheathed lumber diaphragms (see *SDPWS* 4.2.7.1). Apparent shear stiffness in *SDPWS* Table 4.2C is based on assumptions of relative stiffness and nail slip (see C4.2.2).

# C4.3 Wood Shear Walls

## C4.3.1 Application Requirements

General requirements for wood shear walls include consideration of shear wall deflection (discussed in 4.3.2) and strength (discussed in 4.3.3).

## C4.3.2 Deflection

The deflection of a shear wall can be calculated by summing the effects of four sources of deflection: framing bending deflection, panel shear deflection, deflection from nail slip, and deflection due to tie-down slip:

$$\delta_{SW} = \underbrace{\frac{8vh^3}{EAb}}_{(bending)} + \underbrace{\frac{vh}{G_v t_v}}_{(shear)} + \underbrace{0.75he_n}_{(nail\ slip)} + \underbrace{\frac{h}{b}\Delta_a}_{(tie\text{-}down\ nail\ slip)} \quad \text{(C4.3.2-1)}$$

**where:**

v = induced unit shear, plf

h = shear wall height

E = modulus of elasticity of end posts, psi

A = area of end posts cross-section, $in.^2$

b = shear wall length

$G_v t_v$ = shear stiffness, lb/in. of panel depth. See Table C4.2.2A or C4.2.2B.

$\Delta_a$ = total vertical elongation of wall anchorage system (including fastener slip, device elongation, rod elongation, etc.) at the induced unit shear in the shear wall, in.

$e_n$ = nail slip, in. See Table C4.2.2D.

*SDPWS* Equation 4.3-1 is a simplification of Equation C4.3.2-1, using only three terms for calculation of shear wall deflection:

$$\delta_{sw} = \underbrace{\frac{8vh^3}{EAb}}_{(bending)} + \underbrace{\frac{vh}{1000G_a}}_{(shear)} + \underbrace{\frac{h}{b}\Delta_a}_{(tie\text{-}down\ nail\ slip)} \quad \text{(C4.3.2-2)}$$

**where:**

v = induced unit shear, plf

h = shear wall height

E = modulus of elasticity of end posts, psi

A = area of end post cross-section, $in.^2$

b = shear wall length

$G_a$ = apparent shear wall shear stiffness, kips/in.

$\Delta_a$ = total vertical elongation of wall anchorage system (including fastener slip, device elongation, rod elongation, etc.) at the induced unit shear in the shear wall, in.

In *SDPWS* Equation 4.3-1, deflection due to panel shear and nail slip are accounted for by a single apparent shear stiffness term, $G_a$. Calculated deflection, using either the 4-term (Equation C4.3.2-1) or 3-term equation (*SDPWD* Equation 4.3-1), are identical at 1.4 times the allowable shear value for seismic (see Figure C4.3.2). Small "absolute" differences in calculated deflection, below 1.4 times the allowable shear value for seismic, are generally negligible for design purposes. These small differences, however, can influence load distribution assumptions based on relative stiffness if both deflection calculation methods are used in a design. For consistency and to minimize calculation-based differences, either the 4-term equation or 3-term equation should be used.

Each term of the 3-term deflection equation accounts for independent deflection components that contribute to overall shear wall deflection. For example, apparent shear stiffness is intended to represent only the shear component of deflection and does not also attempt to account for bending or tie-down slip. In many cases, such as for gypsum wallboard shear walls and fiberboard shear walls, results from prior testing (17, 23) used to verify apparent shear stiffness estimates were based on ASTM E72 where effect of bending and tie-down slip are minimized due to the presence of metal tie-down rods in the standard test set-up. The relative contribution of each of the deflection components will vary by aspect ratio of the shear wall. For other than narrow shear walls, deformation due to shear deformation (combined effect of nail slip and panel shear deformation) is the most dominant factor.

The effect of tie-down slip becomes more significant as the aspect ratio increases. The *SDPWS* requires an anchoring device (see *SDPWS* 4.3.6.4.2) at each end of the shear wall where dead load stabilizing moment is not sufficient to prevent uplift due to overturning. For standard anchoring devices (tie-downs), the manufacturer's literature typically includes ASD capacity (based on short-term load duration for wind and seismic), and corresponding deflection of the device at ASD levels. Deflection of the device at strength level forces may also be obtained from manufacturer's literature. Reported deflection may or may not include total deflection of the device relative to the wood post and elongation of the tie-down bolt in tension. All sources of vertical elongation of the anchoring device, such as slip in the connection of the device to the wood post and elongation of the tie-down rod should be considered when estimating the $\Delta_a$ term in *SDPWS* Equation 4.3-1. Estimates of $\Delta_a$ at strength level forces are needed where evaluation drift in accordance with *ASCE 7* is required.

**Figure C4.3.2 Comparison of 4-Term and 3-Term Deflection Equations**

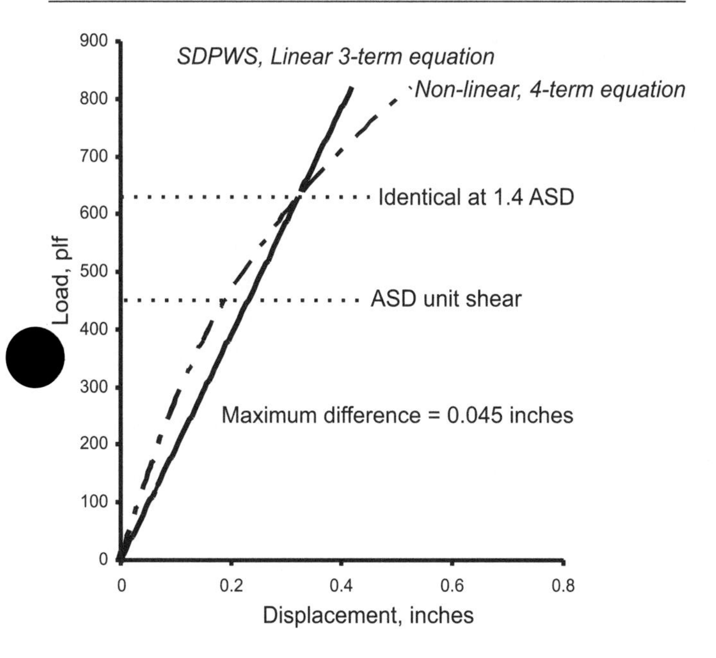


In shear wall table footnotes (*SDPWS* Table 4.3A), a factor of 0.5 is provided to adjust tabulated $G_a$ values (based on fabricated dry condition) to approximate $G_a$ where "green" framing is used. This factor is based on analysis of apparent shear stiffness for wood structural panel shear wall and diaphragm construction where:

1) framing moisture content is greater than 19% at time of fabrication (green), and
2) framing moisture content is less than or equal to 19% at time of fabrication (dry).

The average ratio of "green" to "dry" for $G_a$ across shear wall and diaphragm cells ranged from approximately 0.52 to 0.55. A rounded value of 0.5 results in slightly greater values of calculated deflection for "green" framing when compared to the more detailed 4-term deflection equations. Although based on nail slip relationships applicable to wood structural panel shear walls, this reduction can also be extended to other shear wall types.

In Table C4.3.2A, calculated deflections using *SDPWS* Equation 4.3-1 are compared to deflections from tests at 1.4 times the allowable design value of the assembly for shear walls with fiberboard, gypsum sheathing, and lumber sheathing. Calculated deflections show good agreement (within 1/16") except for cases of horizontal and diagonal lumber sheathing. For lumber sheathing, calculated stiffness is underestimated when compared to test-based stiffness values. However, the lower stated stiffness for horizontal and diagonal lumber sheathing is considered to better reflect stiffness after lumber sheathing dries in service. Early studies (24) suggest that stiffness after drying in service may be ½ of that during tests where friction between boards in lumber sheathed assemblies is a significant factor.

## Table C4.3.2A Data Summary for Fiberboard, Gypsum Wallboard, and Lumber Sheathed Shear Walls

| Reference | Description | Calculated[1] | | | Actual | |
|---|---|---|---|---|---|---|
| | | $1.4v_{s(ASD)}$ (plf) | $G_a$ (kips/in.) | δ (in.) | δ (in.) | $G_a$ (kips/in.) |
| **Fiberboard Sheathing** | | | | | | |
| Ref. 17 | 1/2" fiberboard, roofing nail (11 gage x 1-3/4"), 2" edge spacing, 6" field spacing, 16" stud spacing. 8' x 8' wall. (3 tests). | 364 | 5.5 | 0.53 | 0.46 | 6.3 |
| | 25/32" fiberboard, roofing nail (11 gage x 1-3/4"), 2" edge spacing, 6" field spacing, 16" stud spacing. 8' x 8' wall. (3 tests). | 378 | 5.5 | 0.55 | 0.53 | 5.7 |
| **Gypsum Wallboard (GWB) Sheathing** | | | | | | |
| Ref. 23[2] | 1/2" GWB both sides applied horizontally, GWB Nail (1-1/4") at 8" o.c., 24" stud spacing. 8' x 8' wall. (3 tests). | 184 | 7.0 | 0.21 | 0.17 | 8.7 |
| | 1/2" GWB both sides applied horizontally, GWB Nail (1-1/4") at 8" o.c., 16" stud spacing. 8' x 8' wall. (3 tests). | 245 | 9.6 | 0.20 | 0.16 | 12.2 |
| **Lumber Sheathing** | | | | | | |
| Ref. 24 | Horizontal lumber sheathing. 9' x 14' wall. 1 x 6 and 1 x 8 boards. (2) 8d nails at each stud crossing. Stud spacing 16" o.c. (3 tests - panel 2A, 33, 27). | 70 | 1.5 | 0.42 | 0.25 | 3.9 |
| | Diagonal lumber sheathing (in tension), 9' x 14' wall. 1 x 8 boards. (2) 8d nails at each stud crossing. Stud spacing 16" o.c. (2 tests – panel 5, 31). | 420 | 6.0 | 0.63 | 0.45 | 13.1 |

1. Calculated deflection based on shear component only. For walls tested, small aspect ratio and use of tie-down rods (ASTM E 72) minimize bending and tie-down slip components of deflection.
2. Unit shear and apparent shear stiffness in *SDPWS* Table 4.3B for 7" fastener spacing multiplied by 7/8 to approximate unit shear and stiffness for tested assemblies using 8" fastener spacing.

## EXAMPLE C4.3.2-1 Calculate the Apparent Shear Stiffness, $G_a$, in SDPWS Table 4.3A

Calculate the apparent shear stiffness, $G_a$, in *SDPWS* Table 4.3A for a wood structural panel shear wall constructed as follows:

Sheathing grade = Structural I (OSB)
Common nail size = 6d
Minimum nominal panel thickness = 5/16 in.
Panel edge fastener spacing = 6 in.
Nominal unit shear capacity for seismic, $v_s$ = 400 plf *SDPWS* Table 4.3A

Allowable unit shear capacity for seismic:

$v_{s(ASD)}$ = 400 plf/2 = 200 plf

Panel shear stiffness:

$G_v t_v$ = 77,500 lb/in. of panel depth Table C4.2.2A

Nail load/slip at 1.4 $v_{s(ASD)}$:

$V_n$ = fastener load (lb/nail)
= 1.4 $v_{s(ASD)}$ (6 in.)/(12 in.)
= 140 lb/nail

$e_n = (V_n/456)^{3.144}$ Table C4.2.2D
$= (140/456)^{3.144} = 0.0244$ in.

**Calculate $G_a$:**

$$G_a = \frac{1.4v_{s(ASD)}}{\frac{1.4v_{s(ASD)}}{G_v t_v} + 0.75e_n}$$ Equation C4.2.2-3

$G_a$ = 12,772 lb/in. ≈ 13 kips/in. *SDPWS* Table 4.3A

C4.3.2.1 Deflection of Perforated Shear Walls: The deflection of a perforated shear wall can be calculated using *SDPWS* Equation 4.3-1 using substitution rules as follows to account for the reduced stiffness of full-height perforated shear wall segments:

$v$ = maximum induced unit shear force (plf) in a perforated shear wall per *SDPWS* Equation 4.3-6
b = sum of perforated shear wall segment lengths (full-height), ft

## C4.3.3 Unit Shear Capacities

See C4.2.3 for calculation of ASD allowable unit shear capacity and the LRFD factored unit shear resistance. The shear capacity of perforated shear walls is discussed further in section C4.3.3.4.

C4.3.3.1 Tabulated Nominal Unit Shear Capacities: *SDPWS* Table 4.3A provides nominal unit shear capacities for seismic, $v_s$, and for wind, $v_w$, (see C2.2) for OSB, plywood siding, particleboard, and fiberboard sheathing. *SDPWS* Table 4.3B provides nominal unit shear capacities for gypsum wallboard, gypsum sheathing, plaster, gypsum lath and plaster, and portland cement plaster (stucco). *SDPWS* Appendix Tables A.4.2A, A.4.2B, and A.4.3A provide nominal unit shear capacities for plywood. Nominal unit strength capacities are based on adjustment of allowable values in building codes and industry reference documents (see C2.2).

C4.3.3.2 Summing Shear Capacities: A wall sheathed on two-sides (e.g., a two-sided wall) has twice the capacity of a wall sheathed on one-side (e.g., a one-sided wall) where sheathing material and fastener attachment schedules on each side are identical. Where sheathing materials are the same on both sides, but different fastening schedules are used, provisions of *SDPWS* 4.3.3.2.1 are applicable. Although not common for new construction, use of different fastening schedules is more likely to occur in retrofit of existing construction.

C4.3.3.2.1 For two-sided walls with the same sheathing material on each side (e.g., wood structural panel) and same fastener type, *SDPWS* Equation 4.3-2 and *SDPWS* Equation 4.3-3 provide for determination of the combined stiffness and unit shear capacity based on relative stiffness of each side.

C4.3.3.2.2 For seismic design of two-sided walls with different materials on each side (e.g., gypsum on side one and wood structural panel on side two) the combined unit shear capacity is taken as twice the smaller nominal unit shear capacity or the larger nominal unit shear capacity, whichever is greater. Due to combination rules for seismic design (5), the two-sided unit shear capacity based on different materials on each side of the wall will require use of the least seismic response modification coefficient, R, for calculation of seismic loads. For a two-sided shear wall consisting of wood-structural panel exterior and gypsum wallboard interior, R = 2 is applicable where shear wall design is based on the combined capacity of both sides because R = 2 (associated with gypsum wallboard shear walls in a bearing wall system) is the least R contributing to the two-sided shear wall design capacity. For the same wall condition, when design is based on wood structural panel shear wall capacity alone, R = 6.5 (associated with wood structural panel shear walls in a bearing wall system) is applicable.

For wind design, direct summing of the contribution of gypsum wallboard with the unit shear capacity of wood structural panel, fiberboard, or hardboard panel siding is permitted based on tests (10, 15).

C4.3.3.4 Shear Capacity of Perforated Shear Walls: The shear capacity adjustment factors for perforated shear walls account for the reduced shear wall capacity based on the presence of openings. Shear capacity adjustment factors are based on empirical Equation C4.3.3.4-1. The shear capacity ratio, F, relates the ratio of the shear capacity for a wall with openings to the shear capacity of a fully sheathed wall (13):

$$F = r/(3 - 2r) \qquad (C4.3.3.4\text{-}1)$$

$$r = 1/(1+A_o/(h\sum L_i)) \qquad (C4.3.3.4\text{-}2)$$

**where:**

r = sheathing area ratio

$A_o$ = total area of openings

h = wall height

$\sum L_i$ = sum of the width of full-height sheathing

Agreement between Equation C4.3.3.4-1 and tabulated shear capacity adjustment factors is achieved by recognizing that the tabulated shear capacity adjustment factors are: 1) derived based on an assumption that the height of all openings in a wall are equal to the maximum opening height; and, 2) applied to the sum of the widths of the shear wall segments meeting applicable height-to-width ratios.

## C4.3.5 Shear Wall Types

C4.3.5.1 Segmented Shear Walls: Aspect ratio limits for segmented shear walls are applicable to each full-height segment as shown in Figure C4.3.5.1.

**Figure C4.3.5.1 Typical Segmented Shear Wall Height-to-Width Ratio**

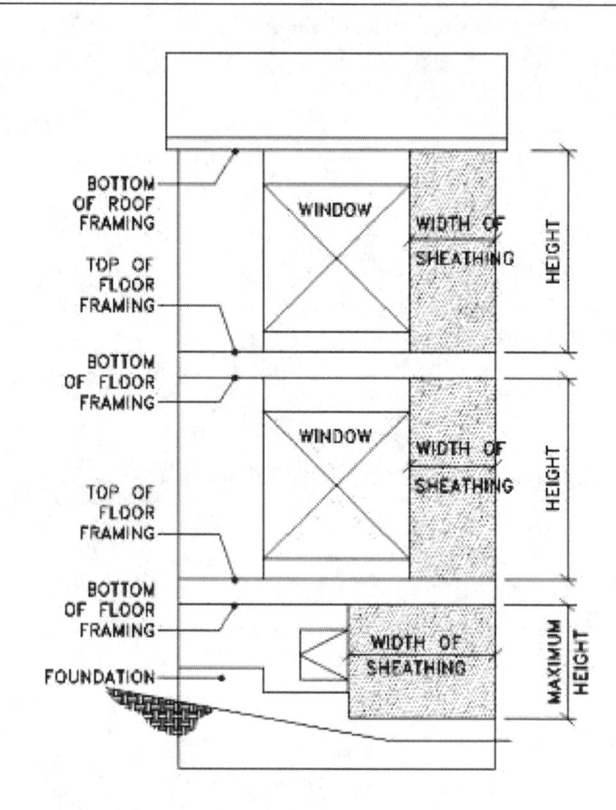


C4.3.5.2 Force Transfer Around Openings: Aspect ratio limits for shear walls designed for force transfer around the opening are applicable to the 1) overall shear wall, and 2) to each wall pier at sides of openings as shown in Figure C4.3.5.2. In addition, the length of the wall pier shall not be less than 2 feet.

### Figure C4.3.5.2 Typical Shear Wall Height-to-Width Ratio for Shear Walls Designed for Force Transfer Around Openings

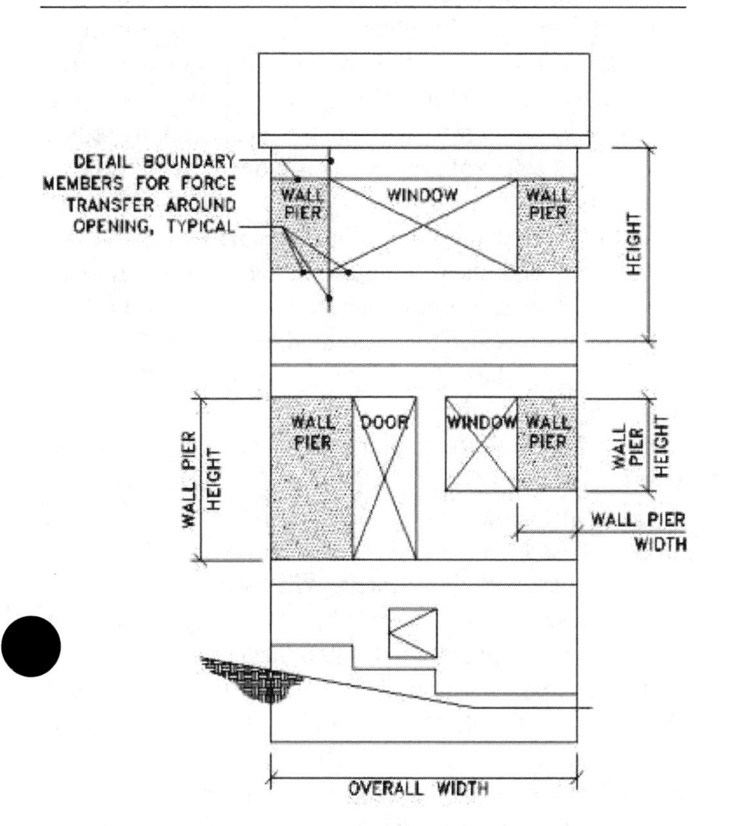


C4.3.5.3 Perforated Shear Walls: For perforated shear walls, aspect ratio limits of *SDPWS* 4.3.4 are applied to full-height wall segments within the perforated shear wall. Full-height sheathed portions within the wall can be designated as perforated shear wall segments where aspect ratio limits of *SDPWS* 4.3.4 are met (see Figure C4.3.5.3).

Perforated shear wall design provisions are applicable to walls with wood structural panel sheathing designed and constructed in accordance with provisions as outlined in *SDPWS* 4.3.5.3. Limits on shear capacity are given in terms of nominal unit strength for single-sided and double-sided perforated shear walls. For single-sided walls, the nominal unit shear capacity shall not exceed 980 plf for seismic or 1,370 plf for wind. For double-sided walls, the nominal unit shear capacity shall not exceed two times 980 plf (or 1,960 plf) for seismic or 2,000 plf for wind. The double-sided limit on nominal unit shear capacity for wind is not two times the single-sided value (e.g., 2 x 1,370 plf = 2,740 plf) because testing of double-sided walls has been limited to 2,000 plf (15).

### Figure C4.3.5.3 Typical Shear Wall Height-to-Width Ratio for Perforated Shear Walls

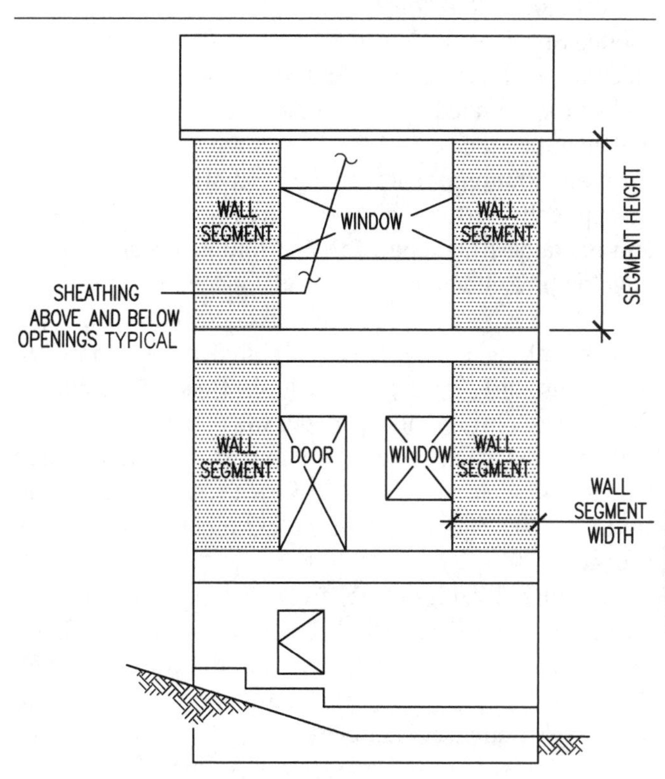


Anchorage and load path requirements for perforated shear walls are specified in *SDPWS* 4.3.6.1.2, 4.3.6.4.1.1, 4.3.6.4.2.1, and 4.3.6.4.4. Anchorage for uplift at perforated shear wall ends, shear, uplift between perforated shear wall ends, and compression chord forces are prescribed to address the non-uniform distribution of shear within a perforated shear wall (7). Prescribed forces for shear and uplift connections are intended to be in excess of the capacity of the individual wall segments such that wall capacity based on the sheathing to framing attachment (shear wall nailing) is not limited by bottom plate attachment for shear and/or uplift.

## C4.3.6 Construction Requirements

C4.3.6.1 Framing Requirements: Framing requirements are intended to ensure that boundary members and other framing are adequately sized to resist induced loads.

C4.3.6.1.1 Tension and Compression Chords: *SDPWS* Equation 4.3-4 provides for calculation of tension and compression chord force due to induced unit shear acting at the top of the wall (e.g., tension and compression due to wall overturning moment). To provide an adequate load path per *SDPWS* 4.3.6.4.4, design of elements and connections must consider forces contributed by each story (i.e., shear and overturning moment must be accumulated and accounted for in the design).

C4.3.6.1.2 Tension and Compression Chords of Perforated Shear Walls: *SDPWS* Equation 4.3-5 provides for calculation of tension force and compression force at each end of a perforated shear wall, due to shear in the wall, and includes the term $1/C_o$ to account for the non-uniform distribution of shear in a perforated shear wall. For example, the perforated shear wall segment with tension end restraint at the end of the perforated shear wall can develop the segmented shear wall capacity (7).

C4.3.6.3.1 Adhesives: Adhesive attachment of shear wall sheathing is generally prohibited unless approved by the authority having jurisdiction. Because of limited ductility and brittle failure modes of rigid adhesive shear wall systems, such systems are limited to seismic design categories A, B, and C and the values of R and $\Omega_0$ are limited (R =1.5 and $\Omega_0$ = 2.5 unless other values are approved).

Tabulated values of apparent shear stiffness, $G_a$, are based on assumed nail slip behavior (see Table C4.2.2D) and are therefore not applicable for adhesive shear wall systems where shear wall sheathing is rigidly bonded to shear wall boundary members. Consideration should be given to increased stiffness where adhesives are used (see C4.1.3 and C4.2.5).

C4.3.6.4.1.1 In-plane Shear Anchorage for Perforated Shear Walls: *SDPWS* Equation 4.3-6 for in-plane shear anchorage includes the term $1/C_o$ to account for the non-uniform distribution of shear in a perforated shear wall. For example, the perforated shear wall segment with tension end restraint at the end of the perforated shear wall can develop the segmented shear wall capacity (7).

C4.3.6.4.2.1 Uplift Anchorage for Perforated Shear Walls: Attachment of the perforated shear wall bottom plate to elements below is intended to ensure that the capacity of the wall is governed by the sheathing to framing attachment (shear wall nailing) and not bottom plate attachment for shear (see C4.3.6.4.1.1) and uplift. An example design (7) provides typical details for transfer of uplift forces.

C4.3.6.4.3 Anchor Bolts: Plate washer size and location are specified for anchoring of wall bottom plates to minimize potential for cross-grain bending failure in the bottom plate. For a 2-1/2" x 2-1/2" plate washer centered on the wide face of a 2x4 bottom plate, edges of the plate washer are within 1/2" of both edges of the wall. For wider bottom plates, such as 2x6, a larger plate washer may be used so that the edge of the plate washer extends to within 1/2" of the sheathed side, or alternatively, the anchor bolt can be located such that the 2-1/2" x 2-1/2" plate washer extends to with 1/2" of the sheathed side of the wall.

C4.3.6.4.4 Load Path: Specified requirements for shear, tension, and compression in *SDPWS* 4.3.6 are to address the effect of induced unit shear on individual wall elements. Overall design of an element must consider forces contributed from multiple stories (i.e., shear and moment must be accumulated and accounted for in the design). In some cases, the presence of load from stories above may increase forces (e.g., effect of gravity loads on compression end posts) while in other cases it may reduce forces (e.g., effect of gravity loads reduces net tension on end posts).

Consistent with a continuous load path for segmented shear walls and shear walls designed for force transfer around openings, a continuous load path to the foundation must also be provided for perforated shear walls. Consideration of accumulated forces (for example, from the stories above) is required and may lead to increases or decreases in member/connection requirements. Accumulation of forces will affect tie-downs at each end of the perforated shear wall, compression resistance at each end of each perforated shear wall segment, and distributed forces v and *t* at each perforated shear wall segment. Where ends of perforated shear wall segments occur over beams or headers, the beam or header will need to be checked for the vertical tension and compression forces in addition to gravity forces. Where adequate collectors are provided to distribute shear, the average shear in the perforated shear wall above (e.g., equivalent to design shear loads), and not the increased shear for anchorage of upper story wall bottom plates to elements below (7), needs to be considered.

## C4.3.7 Shear Wall Systems

Requirements for shear wall sheathing materials, framing, and nailing are consistent with industry recommendations and building code requirements. Minimum framing thickness for all shear wall types is 2" nominal with maximum spacing between framing of 24". Edges of wood-based panels (wood structural panel, particleboard, and fiberboard) are required to be backed by blocking or framing. In addition, fasteners are to be placed at least 3/8" from edges and ends of panels but not less than distances specified by the manufacturer in the manufacturer's literature or code evaluation report.

C4.3.7.1 Wood Structural Panel Shear Walls: For wood structural panel shear walls, framing members or blocking is required at the edges of all panels and a minimum panel dimension of 4' x 8' is specified except at boundaries and changes in framing. Wall construction is intended to consist primarily of full-size sheets except where wall dimensions require use of smaller sheathing pieces. Less than full size pieces of sheathing do not significantly affect wall strength and stiffness (14).

A single 3x nominal framing member is specified at adjoining panel edges for cases prone to splitting and where nominal unit shear capacity exceeds 700 plf in seismic design categories (SDC) D, E, and F. An alternative to single 3x nominal framing, based on principles of mechanics, is use of 2-2x members adequately fastened together. Test results (22, 25) confirm that performance is comparable to a single 3x member. The attachment of the 2-2x members to each other should equal or exceed design unit shear forces in the shear wall. An alternative, capacity-based approach, considers the capacity of the sheathing to framing connection at the adjoining panel edge such that the connection between the 2-2x members equals or exceeds the capacity of the sheathing to framing attachment.

C4.3.7.2 Particleboard Shear Walls: Panel size requirements are consistent with those for wood structural panels (see C4.3.7.1). Apparent shear stiffness in *SDPWS* Table 4.3A is based on assumptions of relative stiffness and nail slip (see C4.2.2 and C4.3.2).

C4.3.7.3 Fiberboard Shear Walls: Panel size requirements are consistent with those for wood structural panels (see C4.3.7.1). Apparent shear stiffness in *SDPWS* Table 4.3A is based on assumptions of relative stiffness and nail slip (see C4.2.2 and C4.3.2).

C4.3.7.4 Gypsum Wallboard, Gypsum Veneer Base, Water-Resistant Backing Board, Gypsum Sheathing, Gypsum Lath and Plaster, or Portland Cement Plaster Shear Walls: The variety of gypsum-based sheathing materials reflects systems addressed in the model building code (2). Appropriate use of these systems requires adherence to referenced standards for proper materials and installation. Where gypsum wallboard is used as a shear wall, edge fastening (e.g., nails or screws) in accordance with *SDPWS* Table 4.3B requirements should be specified and overturning restraint provided where applicable (see *SDPWS* 4.3.6.4.2). Apparent shear stiffness in *SDPWS* Table 4.3B is based on assumptions of relative stiffness and nail slip (see C4.2.2 and C4.3.2).

C4.3.7.5 Shear Walls Diagonally Sheathed with Single-Layer of Lumber: Diagonally sheathed lumber shear walls have comparable strength and stiffness to many wood structural panel shear wall systems. Apparent shear stiffness in *SDPWS* Table 4.3C is based on assumptions of relative stiffness and nail slip (see C4.2.2 and C4.3.2). Early reports (24) indicated that diagonally sheathed lumber shear walls averaged four times the rigidity of horizontally sheathed lumber walls when the boards were loaded primarily in tension. Where load was primarily in compression, a single test showed about seven times the rigidity of a horizontally sheathed lumber wall.

C4.3.7.6 Shear Walls Diagonally Sheathed with Double-Layer of Lumber: Double diagonally sheathed lumber shear walls have comparable strength and stiffness to many wood structural panel shear wall systems. Apparent shear stiffness in *SDPWS* Table 4.3C is based on assumptions of relative stiffness and nail slip (see C4.2.2 and C4.3.2).

C4.3.7.7 Shear Walls Horizontally Sheathed with Single-Layer of Lumber: Horizontally sheathed lumber shear walls have limited unit shear capacity and stiffness when compared to those provided by wood structural panel shear walls of the same overall dimensions. Early reports (21, 24) attributed strength and stiffness of lumber sheathed walls to nail couples at stud crossings and verified low unit shear capacity and stiffness when compared to other bracing methods.

# APPENDIX A

Tabulated apparent shear stiffness values, $G_a$, for plywood can be derived using the same method as shown for OSB sheathing where the sheathing shear stiffness, $G_v t_v$, for plywood is taken from Table C4.2.2A (see C4.3.2 and C4.3.3 for effect of framing moisture content on apparent shear stiffness).

Tabulated values of $G_a$ are based on 3-ply plywood. Separate values of $G_a$ for 4-ply, 5-ply, and COM-PLY were calculated and ratios of these values to $G_a$ based on 3-ply were shown to be in the order of 1.09 to 1.22 for shear walls and 1.04 to 1.16 for diaphragms. Given this relatively small variance, a $G_a$ multiplier of 1.2 was conservatively chosen for 4-ply, 5-ply, and COM-PLY in table footnotes. This option was considered preferable to tabulating $G_a$ values for 3-ply, 4-ply, 5-ply, and COM-PLY separately.

# REFERENCES


1. ASTM Standard D6555-03, Standard Guide for Evaluating System Effects in Repetitive-Member Wood Assemblies, ASTM, West Conshohocken, PA, 2003.

2. International Building Code (IBC), International Code Council, Falls Church, VA, 2006.

3. Laboratory Report 55, Lateral Tests On Plywood Sheathed Diaphragms (out of print), Douglas Fir Plywood Association (now APA – The Engineered Wood Association), Tacoma, WA, 1952.

4. Laboratory Report 63a, 1954 Horizontal Plywood Diaphragm Tests (out of print), Douglas Fir Plywood Association (now APA – The Engineered Wood Association), Tacoma, WA, 1955.

5. Minimum Design Loads for Buildings and Other Structures, American Society of Civil Engineers, ASCE/SEI Standard 7-05, Reston, VA, 2006.

6. National Design Specification (NDS) for Wood Construction, ANSI/AF&PA NDS-2005, American Forest & Paper Association, Washington, DC, 2005.

7. NEHRP Recommended Provisions for Seismic Regulations for New Buildings and Other Structures and Commentary, FEMA Report 450-1 and 2, 2003 Edition, Washington, DC, 2004.

8. Performance Standard for Wood-Based Structural Use Panels, DOC PS2-92, United States Department of Commerce, National Institute of Standards and Technology, Gaithersburg, MD, 1992.

9. Polensek, Anton, Rational Design Procedure for Wood-Stud Walls Under Bending and Compression, Wood Science, July 1976.

10. Racking Load Tests for the American Fiberboard Association and the American Hardboard Association, PFS Test Report #01-25, Madison, WI, 2001.

11. Report 106, 1966 Horizontal Plywood Diaphragm Tests (out of print), Douglas Fir Plywood Association (now APA – The Engineered Wood Association), APA, Tacoma, WA, 1966.

12. Ryan, T.J., Fridley, K.J., Pollock, D.G., and Itani, R.Y., Inter-Story Shear Transfer in Woodframe Buildings: Final Report, Washington State University, Pullman, WA, 2001.

13. Sugiyama, Hideo, The Evaluation of Shear Strength of Plywood-Sheathed Walls with Openings, Mokuzai Kogyo (Wood Industry) 36-7, 1981.

14. Using Narrow Pieces of Wood Structural Panel Sheathing in Wood Shear Walls, APA T2005-08, APA – The Engineered Wood Association, Tacoma, WA, 2005.

15. Wood Structural Panel Shear Walls with Gypsum Wallboard and Window/Door Openings, APA 157, APA – The Engineered Wood Association, Tacoma, WA, 1996.

16. ASTM Standard D5457-04, Standard Specification for Computing the Reference Resistance of Wood-Based Materials and Structural Connections for Load and Resistance Factor Design, ASTM, West Conshohocken, PA, 2004.

17. Racking Load Tests for the American Fiberboard Association, PFS Test Report #96-60, Madison, WI, 1996.

18. Wood Structural Panel Shear Walls, Research Report 154, APA – The Engineered Wood Association, Tacoma, WA, 1993.

19. Plywood Diaphragms, Research Report 138, APA – The Engineered Wood Association, Tacoma, WA, 1990.

20. Wood Frame Construction Manual (WFCM) for One- and Two-Family Dwellings, ANSI/AF&PA WFCM-2001, American Forest & Paper Association, Washington, DC, 2001.

21. Luxford, R.F., Bonser, W.E., Adequacy of Light Frame Wall Construction, No. 2137, Madison, WI: U.S. Department of Agriculture, Forest Service, Forest Products Laboratory, 1958.

22. Shear Wall Lumber Framing: 2x's vs. Single 3x's at Adjoining Panel Edges, APA Report T2003-22, APA – The Engineered Wood Association, Tacoma, WA, 2003.

23. Racking strengths and stiffnesses of exterior and interior frame wall constructions for Department of Housing and Urban Development, Washington, D.C. NAHB Research Foundation, Inc., May, 1971.

24. The Rigidity and Strength of Frame Walls, No. 896. Madison, WI: U.S. Department of Agriculture, Forest Service, Forest Products Laboratory, March 1956.

25. Rosowsky, D., Elkins, L., Carroll, C., Cyclic tests of engineered shear walls considering different plate washer sizes, Oregon State University, Corvallis, OR, 2004.

26. Stillinger, J.R., Lateral Tests on Full-scale Lumber-sheathed Roof Diaphragms, Report No.T-6, Oregon State University, Corvallis, OR, 1953.